AS Maths

OCR Core 1

AS level maths is seriously tricky — no question about that.

We've done everything we can to make things easier for you.
We've scrutinised past paper questions and we've gone through the syllabuses with a fine-toothed comb. So we've found out exactly what you need to know, then explained it simply and clearly.

We've stuck in as many helpful hints as you could possibly want
— then we even tried to put some funny bits in to keep you awake.

We've done our bit — the rest is up to you.

What CGP is all about

Our sole aim here at CGP is to produce the highest quality books — carefully written, immaculately presented and dangerously close to being funny.

Then we work our socks off to get them out to you
— at the cheapest possible prices.

Contents

SECTION 5 — DIFFERENTIATION

P1 — PRACTICE EXAM 1

P1 — PRACTICE EXAM 2

This book covers the Core 1 module of the OCR specification.

Published by Coordination Group Publications Ltd.

Contributors:
Charley Darbishire
Simon Little
Andy Park
Glenn Rogers
Mike Smith
Claire Thompson
Kieran Wardell

And:
Iain Nash

Updated by:
Andy Park
Tim Major
Alice Shepperson
Claire Thompson
Julie Wakeling

With thanks to Claire Baldwin and Colin Wells for the proofreading.

ISBN: 978 1 84146 763 4

Groovy website: www.cgpbooks.co.uk
Jolly bits of clipart from CorelDRAW®
Printed by Elanders Ltd, Newcastle upon Tyne.

Based on the classic CGP style created by Richard Parsons.

A Few Definitions and Things

Yep, this is a pretty dull way to start a book. A list of definitions. But in the words of that annoying one-hit wonder bloke in the tartan suit, things can only get better. Which is nice.

Polynomials

POLYNOMIALS are expressions of the form $a + bx + cx^2 + dx^3 + ...$

$5y^3 + 2y + 23$ ← Polynomial in the variable y.

$1 + x^2$

$z^{42} + 3z - z^2 - 1$ ← Polynomial in the variable z.

The bits separated by the +/– signs are <u>terms</u>.

x, y and z are always VARIABLES. They're usually what you solve equations to find. They often have more than one possible value.

Letters like a, b, c are always CONSTANTS. Constants never change. They're fixed numbers — but can be represented by letters. π is a good example. You use the symbol π, but it's just a number = 3.1415...

Functions

FUNCTIONS take a value, do something to it, and output another value.

$f(x) = x^2 + 1$ ← function f takes a value, squares it and adds 1.

$g(x) = 2 - \sin 2x$ ← function g takes a value (in degrees), doubles it, takes the sine of it, then takes the value away from 2.

You can plug values into a function — just replace the variable with a certain number.

$f(-2) = (-2)^2 + 1 = 5$

$f(0) = (0)^2 + 1 = 1$

$f(252) = (252)^2 + 1 = 63505$

$g(-90) = 2 - \sin(-180°) = 2 - 0 = 2$

$g(0) = 2 - \sin 0° = 2 - 0 = 2$

$g(45) = 2 - \sin 90° = 2 - 1 = 1$

Exam questions use functions all the time. They generally don't have that much to do with the actual question. It's just a bit of terminology to get comfortable with.

Multiplication and Division

There's three different ways of showing MULTIPLICATION:

1) with good old-fashioned "times" signs (×):

$f(x) = (2x \times 6y) + (2x \times \sin x) + (z \times y)$

The multiplication signs and the variable x are easily confused.

2) or sometimes just use a little dot:

$f(x) = 2x.6y + 2x.\sin x + z.y$

Dots are better for long expressions — they're less confusing and easier to read.

3) but you often don't need anything at all:

$f(x) = 12xy + 2x \sin x + zy$

And there's three different ways of showing DIVISION:

1) $\dfrac{x+2}{3}$

2) $(x+2) \div 3$

3) $(x+2)/3$

Equations and Identities

This is an IDENTITY:

$x^2 - y^2 \equiv (x + y)(x - y)$

Make up any values you like for x and y, and it's always true. The left-hand side always equals the right-hand side.

But this is an EQUATION:

$y = x^2 + x$

This has at most two possible solutions for each value of y. e.g. if y=0, x can only be 0 or -1.

The difference is that the identity's true for <u>all</u> values of x and y, but the equation's only true for certain values.

NB: If it's an identity, use the $\equiv$ sign instead of =.

Laws of Indices

You use the laws of indices a helluva lot in maths — when you're integrating, differentiating and ...er... well loads of other places. So take the time to get them sorted <u>now</u>.

Three mega-important **Laws of Indices**

You <u>must</u> know these three rules. I can't make it any clearer than that.

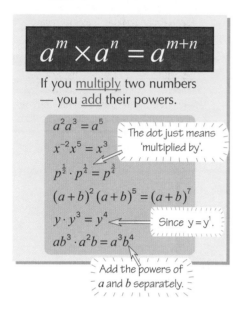

$$a^m \times a^n = a^{m+n}$$

If you <u>multiply</u> two numbers — you <u>add</u> their powers.

$a^2 a^3 = a^5$

$x^{-2} x^5 = x^3$

$p^{\frac{1}{2}} \cdot p^{\frac{1}{4}} = p^{\frac{3}{4}}$

The dot just means 'multiplied by'.

$(a+b)^2 (a+b)^5 = (a+b)^7$

$y \cdot y^3 = y^4$ — Since $y = y^1$.

$ab^3 \cdot a^2 b = a^3 b^4$

Add the powers of a and b separately.

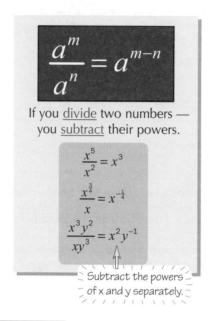

$$\frac{a^m}{a^n} = a^{m-n}$$

If you <u>divide</u> two numbers — you <u>subtract</u> their powers.

$\dfrac{x^5}{x^2} = x^3$

$\dfrac{x^{\frac{3}{4}}}{x} = x^{-\frac{1}{4}}$

$\dfrac{x^3 y^2}{xy^3} = x^2 y^{-1}$

Subtract the powers of x and y separately.

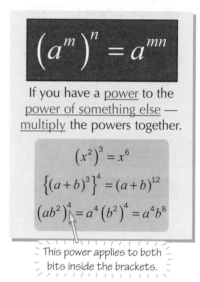

$$\left(a^m\right)^n = a^{mn}$$

If you have a <u>power</u> to the <u>power of something else</u> — <u>multiply</u> the powers together.

$\left(x^2\right)^3 = x^6$

$\left\{(a+b)^3\right\}^4 = (a+b)^{12}$

$\left(ab^2\right)^4 = a^4 \left(b^2\right)^4 = a^4 b^8$

This power applies to both bits inside the brackets.

Other important stuff about **Indices**

You can't get very far without knowing this sort of stuff. Learn it — you'll definitely be able to use it.

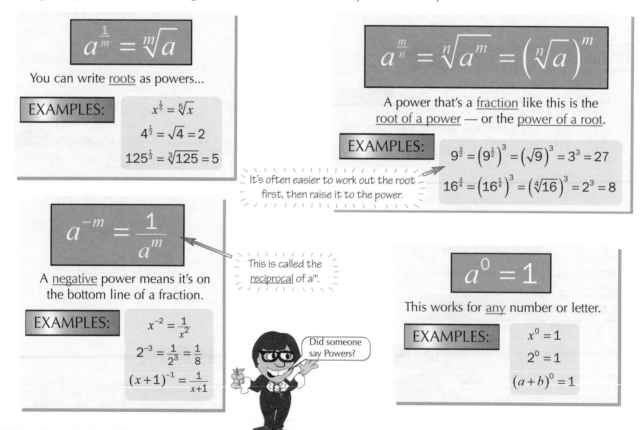

$$a^{\frac{1}{m}} = \sqrt[m]{a}$$

You can write <u>roots</u> as powers...

EXAMPLES:

$x^{\frac{1}{5}} = \sqrt[5]{x}$

$4^{\frac{1}{2}} = \sqrt{4} = 2$

$125^{\frac{1}{3}} = \sqrt[3]{125} = 5$

$$a^{\frac{m}{n}} = \sqrt[n]{a^m} = \left(\sqrt[n]{a}\right)^m$$

A power that's a <u>fraction</u> like this is the <u>root of a power</u> — or the <u>power of a root</u>.

EXAMPLES:

$9^{\frac{3}{2}} = \left(9^{\frac{1}{2}}\right)^3 = \left(\sqrt{9}\right)^3 = 3^3 = 27$

$16^{\frac{3}{4}} = \left(16^{\frac{1}{4}}\right)^3 = \left(\sqrt[4]{16}\right)^3 = 2^3 = 8$

It's often easier to work out the root first, then raise it to the power.

$$a^{-m} = \frac{1}{a^m}$$

This is called the <u>reciprocal</u> of a^m.

A <u>negative</u> power means it's on the bottom line of a fraction.

EXAMPLES:

$x^{-2} = \dfrac{1}{x^2}$

$2^{-3} = \dfrac{1}{2^3} = \dfrac{1}{8}$

$(x+1)^{-1} = \dfrac{1}{x+1}$

Did someone say Powers?

$$a^0 = 1$$

This works for <u>any</u> number or letter.

EXAMPLES:

$x^0 = 1$

$2^0 = 1$

$(a+b)^0 = 1$

Indices, indices — de fish all live indices...

What can I say that I haven't said already? Blah, blah, <u>important</u>. Blah, blah, <u>learn</u> these. Blah, blah, <u>use</u> them all the time. Mmm, that's about all that needs to be said really. So I'll be quiet and let you get on with what you need to do.

Surds

A surd is a number like $\sqrt{2}$, $\sqrt[3]{12}$ or $5\sqrt{3}$ — one that's written with the $\sqrt{}$ sign. They're important because you can give <u>exact</u> answers where you'd otherwise have to round to a certain number of decimal places.

Surds are sometimes the only way to give an Exact Answer

Put $\sqrt{2}$ into a calculator and you'll get something like 1.414213562...
But square 1.414213562 and you get 1.999999999.

And no matter how many decimal places you use, you'll never get <u>exactly</u> 2.
The only way to write the exact, spot on value is to <u>use surds</u>.

There are basically Three Rules for using Surds

There are three <u>rules</u> you'll need to know to be able to use surds properly. Check out the 'Rules of Surds' box below.

EXAMPLES: (i) Simplify $\sqrt{12}$ and $\sqrt{\frac{3}{16}}$.   (ii) Show that $\frac{9}{\sqrt{3}} = 3\sqrt{3}$.   (iii) Find $\left(2\sqrt{5}+3\sqrt{6}\right)^2$.

(i) <u>Simplifying</u> surds means making the number in the $\sqrt{}$ sign <u>smaller</u>, or getting rid of a <u>fraction</u> in the $\sqrt{}$ sign.

$$\sqrt{12}=\sqrt{4\times3}=\sqrt{4}\times\sqrt{3}=2\sqrt{3}$$

$$\sqrt{\frac{3}{16}}=\frac{\sqrt{3}}{\sqrt{16}}=\frac{\sqrt{3}}{4}$$

Using $\sqrt{\frac{a}{b}}=\frac{\sqrt{a}}{\sqrt{b}}$.

Using $\sqrt{ab}=\sqrt{a}\sqrt{b}$.

(ii) For questions like these, you have to write a number (here, it's 3) as $3=\left(\sqrt{3}\right)^2=\sqrt{3}\times\sqrt{3}$.

$$\frac{9}{\sqrt{3}}=\frac{3\times3}{\sqrt{3}}=\frac{3\times\sqrt{3}\times\sqrt{3}}{\sqrt{3}}=3\sqrt{3}$$

Cancelling $\sqrt{3}$ from the top and bottom lines.

(iii) Multiply surds very <u>carefully</u> — it's easy to make a silly mistake.

$$\left(2\sqrt{5}+3\sqrt{6}\right)^2=\left(2\sqrt{5}+3\sqrt{6}\right)\left(2\sqrt{5}+3\sqrt{6}\right)$$
$$=\left(2\sqrt{5}\right)^2+2\times\left(2\sqrt{5}\right)\times\left(3\sqrt{6}\right)+\left(3\sqrt{6}\right)^2$$
$$=\left(2^2\times\sqrt{5}^2\right)+\left(2\times2\times3\times\sqrt{5}\times\sqrt{6}\right)+\left(3^2\times\sqrt{6}^2\right)$$
$$=20+12\sqrt{30}+54$$
$$=74+12\sqrt{30}$$

$=4\times5=20$ $=12\sqrt{5}\sqrt{6}=12\sqrt{30}$ $=9\times6=54$

Rules of Surds

There's not really very much to remember.

$$\sqrt{ab}=\sqrt{a}\sqrt{b}$$
$$\sqrt{\frac{a}{b}}=\frac{\sqrt{a}}{\sqrt{b}}$$
$$a=\left(\sqrt{a}\right)^2=\sqrt{a}\sqrt{a}$$

But don't forget that:
$$\sqrt{a+b}\neq\sqrt{a}+\sqrt{b}$$

Remove surds from the bottom of fractions by Rationalising the Denominator

Surds are pretty darn complicated.
So they're the last thing you want at the bottom of a fraction.
But have no fear — <u>Rationalise the Denominator</u>...
Yup, you heard... (it means getting rid of the surds from the bottom of a fraction).

EXAMPLE: Rationalise the denominator of $\frac{1}{1+\sqrt{2}}$

Multiply the top and bottom by the denominator (but change the sign in front of the surd).

$$\frac{1}{1+\sqrt{2}}\times\frac{1-\sqrt{2}}{1-\sqrt{2}}$$
$$\frac{1-\sqrt{2}}{(1+\sqrt{2})(1-\sqrt{2})}=\frac{1-\sqrt{2}}{1^2+\sqrt{2}-\sqrt{2}-\sqrt{2}^2}$$
$$\frac{1-\sqrt{2}}{1-2}=\frac{1-\sqrt{2}}{-1}=-1+\sqrt{2}$$

This works because: $(a+b)(a-b)=a^2-b^2$

Surely the pun is mightier than the surd...

There's not much to surds really — but they cause a load of hassle. Think of them as just ways to save you the bother of getting your calculator out and pressing buttons — then you might grow to know and love them. The box of rules in the middle is the vital stuff. Learn them till you can write them down without thinking — then get loads of practice with them.

Multiplying Out Brackets

In this horrific nightmare that is AS-level maths, you need to manipulate and simplify expressions all the time.

Remove brackets by Multiplying them out

Here's the basic types you have to deal with. You'll have seen them before. But there's no harm in reminding you, eh?

Multiply Your Brackets Here — we do all shapes and sizes

Single Brackets

$$a(b+c+d) = ab + ac + ad$$

Squared Brackets

$$(a+b)^2 = (a+b)(a+b) = a^2 + 2ab + b^2$$

Use the middle stage until you're comfortable with it. Just *never* make this *mistake*: $(a+b)^2 = a^2 + b^2$

Double Brackets

$$(a+b)(c+d) = ac + ad + bc + bd$$

Long Brackets

Write it out again with each term from one bracket separately multiplied by the other bracket.

$$(x+y+z)(a+b+c+d)$$
$$= x(a+b+c+d) + y(a+b+c+d) + z(a+b+c+d)$$

Then multiply out each of these brackets, one at a time.

Single Brackets

$$3xy(x^2 + 2x - 8)$$

Multiply all the terms inside the brackets by the bit outside — separately.

All the stuff in the brackets now needs sorting out. Work on each bracket separately.

$$(3xy \times x^2) + (3xy \times 2x) + (3xy \times (-8))$$

I've put brackets round each bit to make it easier to read.

$$(3x^3y) + (6x^2y) + (-24xy)$$

Multiply the numbers first, then put the x's and other letters together.

$$3x^3y + 6x^2y - 24xy$$

Squared Brackets

Either write it as two brackets and multiply it out...

$$(2y^2 + 3x)^2$$
$$(2y^2 + 3x)(2y^2 + 3x)$$

The dot just means 'multiplied by' — the same as the × sign.

$$2y^2.2y^2 + 2y^2.3x + 3x.2y^2 + 3x.3x$$

From here on it's simplification — nothing more, nothing less.

$$4y^4 + 6xy^2 + 6xy^2 + 9x^2$$
$$4y^4 + 12xy^2 + 9x^2$$

...or do it in one go.

$$(2y^2)^2 + 2(2y^2)(3x) + (3x)^2$$
$$\text{a}^2 \qquad 2\text{ab} \qquad \text{b}^2$$
$$4y^4 + 12xy^2 + 9x^2$$

Long Brackets

$$(2x^2 + 3x + 6)(4x^3 + 6x^2 + 3)$$

Each term in the first bracket has been multiplied by the second bracket.

$$2x^2(4x^3 + 6x^2 + 3) + 3x(4x^3 + 6x^2 + 3) + 6(4x^3 + 6x^2 + 3)$$

Now multiply out each of these brackets.

$$(8x^5 + 12x^4 + 6x^2) + (12x^4 + 18x^3 + 9x) + (24x^3 + 36x^2 + 18)$$

Then simplify it all...

$$\underline{8x^5 + 24x^4 + 42x^3 + 42x^2 + 9x + 18}$$

Go forth and multiply out brackets...

OK, so this is obvious, but I'll say it anyway — if you've got 3 or more brackets together, multiply them out 2 at a time. Then you'll be turning a really hard problem into two easy ones. You can do that loads in maths. In fact, writing the same thing in different ways is what maths is about. That and sitting in classrooms with tacky 'maths can be fun' posters...

Taking Out Common Factors

Common factors need to be hunted down, and taken outside the brackets. They are a danger to your exam mark.

Spot those **Common Factors**

A bit which is in each term of an expression is a <u>common factor</u>.

Spot Those Common Factors $2x^3z + 4x^2yz + 14x^2y^2z$

Look for any bits that are in each term.

<u>Numbers</u>: there's a common factor of 2 here because 2 divides into 2, 4 and 14.

<u>Variables</u>: there's at least an x^2 in each term and there's a z in each term.

So there's a <u>common factor of $2x^2z$</u> in this expression.

And Take Them Outside a Bracket

If you spot a common factor you can "<u>take it out</u>":

Write the common factor outside a bracket.

$2x^2z(x + 2y + 7y^2)$

...and put what's left of each term inside the bracket:

Afterwards, always <u>multiply back out</u> to check you did it right:

Check by Multiplying Out Again

$2x^2z(x + 2y + 7y^2) = 2x^3z + 4x^2yz + 14x^2y^2z$

But it's not just numbers and variables you need to look for...

Trig Functions: $\sin x \sin y + \cos x \sin y$

This has a common factor of sin y.
So take it out to get...

$\sin y(\sin x + \cos x)$

Brackets: $(y + a)^2(x - a)^3 + (x - a)^2$

$(x - a)^2$ is a common factor
— it comes out to give:

$(x - a)^2\left((y + a)^2(x - a) + 1\right)$

Look for **Common Factors** when **Simplifying Expressions**

EXAMPLE: Simplify... $(x + 1)(x - 2) + (x + 1)^2 - x(x + 1)$

There's an (x+1) factor in each term, so we can <u>take this out as a common factor</u> (hurrah).

$(x + 1)\{(x - 2) + (x + 1) - x\}$

The terms inside the big bracket are the old terms with an (x+1) removed.

At this point you should check that this multiplies out to give the original expression. (You can just do this in your head, if you trust it.)

Then simplify the big bracket's innards:

$(x + 1)(x - 2 + x + 1 - x)$

$= (x + 1)(x - 1)$

$= x^2 - 1$

Get this answer by multiplying out the two brackets (or by using the "difference of two squares").

Bored of spotting trains or birds? Try common factors...

You'll be doing this business of taking out common factors a lot — so get your head round this. It's just a case of looking for things that are in all the different terms of an expression, i.e. bits they have in common. And if something's in all the different terms, save yourself some time and ink, and write it once — instead of two, three or more times.

SECTION ONE — ALGEBRA BASICS

Algebraic Fractions

No one likes fractions. But just like Mondays, you can't put them off forever. Face those fears. Here goes...

The first thing you've got to know about fractions:

$$\frac{a}{x} + \frac{b}{x} + \frac{c}{x} \equiv \frac{a+b+c}{x}$$

You can just add the stuff on the top lines because the bottom lines are all the same.

x is called a common denominator — a fancy way of saying 'the bottom line of all the fractions is x'.

Add fractions by putting them over a Common Denominator...

Finding a common denominator just means 'rewriting some fractions so all their bottom lines are the same'.

EXAMPLE: Simplify $\dfrac{1}{2x} - \dfrac{1}{3x} + \dfrac{1}{5x}$

You need to rewrite these so that all the bottom lines are equal. What you want is something that all these bottom lines divide into.

Put It over a Common Denominator

30 is the lowest number that 2, 3, and 5 go into. So the common denominator is 30x.

$$\frac{15}{30x} - \frac{10}{30x} + \frac{6}{30x}$$

Always check that these divide out to give what you started with.

$$= \frac{15-10+6}{30x} = \frac{11}{30x}$$

...even horrible looking ones

Yep, finding a common denominator even works for those fraction nasties — like these:

EXAMPLE: Find $\dfrac{2y}{x(x+3)} + \dfrac{1}{y^2(x+3)} - \dfrac{x}{y}$

Find the Common Denominator

Take all the individual 'bits' from the bottom lines and multiply them together. Only use each bit once unless something on the bottom line is squared.

The individual 'bits' here are x, (x+3) and y...

$$xy^2(x+3)$$

...but you need to use y^2 because there's a y^2 in the second fraction's denominator.

Put Each Fraction over the Common Denominator

Make the denominator of each fraction into the common denominator.

$$\frac{y^2 \times 2y}{y^2 x(x+3)} + \frac{x \times 1}{xy^2(x+3)} - \frac{xy(x+3) \times x}{xy(x+3)y}$$

Multiply the top and bottom lines of each fraction by whatever makes the bottom line the same as the common denominator.

Combine into One Fraction

Once everything's over the common denominator — you can just add the top lines together.

$$= \frac{2y^3 + x - x^2 y(x+3)}{xy^2(x+3)}$$

As always — if you see a minus sign, look out for possible problems.

All the bottom lines are the same — so you can just add the top lines.

$$= \frac{2y^3 + x - x^3 y - 3x^2 y}{xy^2(x+3)}$$

All you need to do now is tidy up the top.

Not the nicest of answers. But it *is* the answer, so it'll have to do.

Well put me over a common denominator and pickle my walrus...

Adding fractions — turning lots of fractions into one fraction. Sounds pretty good to me, since it means you don't have to write as much. Better do it carefully, though — otherwise you can watch those marks shoot straight down the toilet.

Simplifying Expressions

I know this is basic stuff but if you don't get really comfortable with it you <u>will</u> make silly mistakes. You will.

Cancelling stuff on the top and bottom lines

Cancelling stuff is good — because it means you've got rid of something, and you don't have to write as much.

EXAMPLE: Simplify $\dfrac{ax+ay}{az}$

You can do this in two ways. Use whichever you prefer — but make sure you understand the ideas behind both.

Factorise — then Cancel

$$\frac{ax+ay}{az} = \frac{a(x+y)}{az}$$

Factorise the top line.

Cancel the 'a'.

$$= \frac{\cancel{a}(x+y)}{\cancel{a}z} = \frac{x+y}{z}$$

Split into Two Fractions — then Cancel

$$\frac{ax+ay}{az} = \frac{ax}{az} + \frac{ay}{az}$$

This is an okay thing to do — just think what you'd get if you added these.

$$= \frac{\cancel{a}x}{\cancel{a}z} + \frac{\cancel{a}y}{\cancel{a}z} = \frac{x}{z} + \frac{y}{z}$$

This answer's the same as the one from the first box — honest. Check it yourself by adding the fractions.

Simplifying complicated-looking Brackets

EXAMPLE: Simplify the expression $(x-y)(x^2+xy+y^2)$

There's only one thing to do here.... Multiply out those brackets!

$$(x-y)(x^2+xy+y^2) = x(x^2+xy+y^2) - y(x^2+xy+y^2)$$

Multiplying each term in the first bracket by the second bracket.

$$= (x^3+x^2y+xy^2) - (x^2y+xy^2+y^3)$$

Multiplying out each of these two brackets.

$$= x^3+x^2y+xy^2-x^2y-xy^2-y^3$$

Don't forget these become minus signs because of the minus sign in front of the bracket.

And then the x^2y and the xy^2 terms disappear...

$$= x^3 - y^3$$

Sometimes you just have to do Anything you can think of and Hope...

Sometimes it's not easy to see what you're supposed to do to simplify something.
When this happens — just do anything you can think of and see what 'comes out in the wash'.

EXAMPLE: Simplify $4x + \dfrac{4x}{x+1} - 4(x+1)$

There's nothing obvious to do — so do what you can. Try adding them as fractions...

$$4x + \frac{4x}{x+1} - 4(x+1) = \frac{(x+1)\times 4x}{x+1} + \frac{4x}{x+1} - \frac{(x+1)\times 4(x+1)}{x+1}$$

The common denominator is (x + 1).

$$= \frac{4x^2 + 4x + 4x - 4(x+1)^2}{x+1}$$

Still looks horrible. So work out the brackets — but don't forget the minus signs.

$$= \frac{4x^2 + 4x + 4x - 4x^2 - 8x - 4}{x+1}$$

$$= -\frac{4}{x+1}$$

Aha — everything disappears to leave you with this. And this is definitely simpler than it looked at the start.

Don't look at me like that...

Choose a word, any word at all. Like "Simple". Now stare at it. Keep staring at it. Does it look weird? No?
Stare a bit longer. Now does it look weird? Yes? Why is that? I don't understand.

Section One Revision Questions

So that was the first section. And let's face it, it wasn't that bad. But it's all really important <u>basic stuff</u> that you need to be very comfortable with — otherwise AS-level maths will just become a never ending misery. Anyway, before you get stuck into section two, test yourself with these questions. Go on. If you thought this section was a doddle, you should be able fly through them... (They're also very helpful if you're having trouble sleeping.)

1) Pick out the constants and the variables from the following equations:

   a) $(ax + 6)^2 = 2b + 3$ b) $\sin k\theta = 5$ c) $y = \dfrac{-b \pm \sqrt{b^2 - 4ac}}{2a}$ d) $y = x^2 + ax + 2$

2) What symbol should be used instead of the equals sign in identities?

3) Which of these are identities (i.e. true for all variable values)?

   A $(x + b)(y - b) = xy + b(y - x) - b^2$ B $(2y + x)^2 = 10$

   C $\tan\theta = \dfrac{\sin\theta}{\cos\theta}$ D $a^3 + b^3 = (a + b)(a^2 - ab + b^2)$

4) Number 6 in my all-time top ten functions is: f defined by $f(x) = (x+1)^2/3x$. Find the value of f when...

   a) $x = -4$ b) $x = 0$ c) $x = 3$ d) $x = -1$

5) Simplify these:

   a) $x^3 . x^5$ b) $a^7 . a^8$ c) $\dfrac{x^8}{x^2}$ d) $\left(a^2\right)^4$ e) $\left(xy^2\right) . \left(x^3 yz\right)$ f) $\dfrac{a^2 b^4 c^6}{a^3 b^2 c}$

6) Work out the following:

   a) $16^{\frac{1}{2}}$ b) $8^{\frac{1}{3}}$ c) $16^{\frac{3}{4}}$ d) x^0 e) $49^{-\frac{1}{2}}$

7) Find exact answers to these equations:

   a) $x^2 - 5 = 0$ b) $(x + 2)^2 - 3 = 0$

8) Simplify:

   a) $\sqrt{28}$ b) $\sqrt{\dfrac{5}{36}}$ c) $\sqrt{18}$ d) $\sqrt{\dfrac{9}{16}}$

9) Show that a) $\dfrac{8}{\sqrt{2}} = 4\sqrt{2}$, and b) $\dfrac{\sqrt{2}}{2} = \dfrac{1}{\sqrt{2}}$

10) Find $\left(6\sqrt{3} + 2\sqrt{7}\right)^2$

11) Rationalise the denominator of: $\dfrac{2}{3+\sqrt{7}}$

12) Remove the brackets and simplify the following expressions:

   a) $(a + b)(a - b)$ b) $(a + b)(a + b)$

   c) $35xy + 25y(5y + 7x) - 100y^2$ d) $(x + 3y + 2)(3x + y + 7)$

13) Take out the common factors from the following expressions:

   a) $2x^2 y + axy + 2xy^2 \sin x$ b) $\sin^2 x + \cos^2 x \sin^2 x$ c) $16y + 8yx + 56x$ d) $x(x - 2) + 3(2 - x)$

14) Put the following expressions over a common denominator:

   a) $\dfrac{2x}{3} + \dfrac{y}{12} + \dfrac{x}{5}$ b) $\dfrac{5}{xy^2} - \dfrac{2}{x^2 y}$ c) $\dfrac{1}{x} + \dfrac{x}{x + y} + \dfrac{y}{x - y}$

15) Simplify these expressions:

   a) $\dfrac{2a}{b} - \dfrac{a}{2b}$ b) $\dfrac{2p}{p + q} + \dfrac{2q}{p - q}$ c) "A bird in the hand is worth two in the bush"

Sketching Quadratic Graphs

If a question doesn't seem to make sense, or you can't see how to go about solving a problem, try drawing a <u>graph</u>. It sometimes helps if you can actually <u>see</u> what the problem is, rather than just reading about it.

Sketch the graphs of the following quadratic functions:

① $y = 2x^2 - 4x + 3$

② $y = 8 - 2x - x^2$

Quadratic graphs are **Always** u-shaped or n-shaped

 A The first thing you need to know is whether the graph's going to be u-shaped or n-shaped (upside down). To decide, look at the <u>coefficient of x^2</u>.

$y = 2x^2 - 4x + 3$

The coefficient of x^2 here is <u>positive</u>... ...so the graph's u-shaped. +ve

$y = 8 - 2x - x^2$

The coefficient of x^2 here is <u>negative</u>... ...so the graph's upside down (n-shaped). –ve

B Now find the places where the graph crosses the <u>axes</u> (both the y-axis and the x-axis).

(i) Put x=0 to find where it meets the <u>y-axis</u>.

$y = 2x^2 - 4x + 3$

$y = (2 \times 0^2) - (4 \times 0) + 3$ so $y = 3$

That's where it crosses the y axis

(i) Put x=0.

$y = 8 - 2x - x^2$

$y = 8 - (2 \times 0) - 0^2$ so $y = 8$

(ii) Solve y=0 to find where it meets the <u>x-axis</u>.

$2x^2 - 4x + 3 = 0$

$b^2 - 4ac = -8 < 0$

You could use the formula. But first check $b^2 - 4ac$ to see if y = 0 has any roots.

So it has no solutions, and doesn't cross the x-axis.

For more info, see page 15.

(ii) Solve y=0.

$8 - 2x - x^2 = 0$

$\Rightarrow (2 - x)(x + 4) = 0$

$\Rightarrow x = 2 \ or \ x = -4$

This equation factorises easily...

The minimum or maximum of the graph is always at $x = \frac{-b}{2a}$

The maximum value is <u>halfway</u> between the roots — the graph's symmetrical.

C Finally, find the <u>minimum</u> or <u>maximum</u> (i.e. the <u>vertex</u>).

The maximum value is at $x = -1$

So the maximum is $y = 8 - (2 \times -1) - (-1)^2$

i.e. the graph has a maximum at the point (–1,9).

Since $y = 2(x - 1)^2 + 1$

By <u>completing the square</u> (see page 12).

the minimum value is $y = 1$, which occurs at $x = 1$

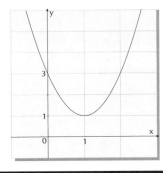

Sketching Quadratic Graphs

A) **up or down** — decide which direction the curve points in.

B) **axes** — find where the curve crosses them.

C) **max / min** — find the turning point.

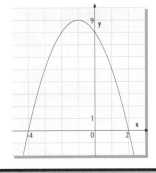

Van Gogh, Monet — all the greats started out sketching graphs...

So there's three steps here to learn. Simple enough. You can do the third step (finding the max/min point) by either a) completing the square, which is covered a bit later, or b) using the fact that the graph's symmetrical — so once you've found the points where it crosses the x-axis, the point halfway between them will be the max/min. It's all laughs here...

Factorising a Quadratic

Factorising a quadratic means putting it into two brackets — and is useful if you're trying to draw a graph of a quadratic or solve a quadratic equation. It's pretty easy if a = 1 (in $ax^2 + bx + c$ form), but can be a real pain otherwise.

$$x^2 - x - 12 = (x - 4)(x + 3)$$

Factorising's not so bad when a = 1

EXAMPLE: Solve $x^2 - 8 = 2x$ by factorising.

{A}

Put into $ax^2 + bx + c = 0$ Form

$x^2 - 2x - 8 = 0$ ← So a = 1, b = –2, c = –8.

Write down the two brackets with x's in: $x^2 - 2x - 8 = (x\ \ \ \ )(x\ \ \ \ )$

{B}

Find the Two Numbers

Find two numbers that <u>multiply</u> together to make 'c' but which also <u>add</u> or <u>subtract</u> to give 'b' (you can ignore any minus signs for now).

1 and 8 multiply to give 8 — and add / subtract to give 9 and 7.

2 and 4 multiply to give 8 — and add / subtract to give 6 and 2.

This is the value for 'b' you're after — so this is the right combination: 2 and 4.

{C}

Find the Signs

Now all you have to do is put in the <u>plus</u> or <u>minus</u> signs.

$x^2 - 2x - 8 = (x\ \ 4)(x\ \ 2)$

If c is negative, then the signs must be different.

$x^2 - 2x - 8 = (x + 2)(x - 4)$

It must be +2 and –4 because 2×(–4)=–8 and 2+(–4)=2–4=–2

{D}

Solve the Equation

All you've done so far is to factorise the equation — you've still got to solve it.

$(x + 2)(x - 4) = 0$

Don't forget this last step. The factors aren't the answer.

$\Rightarrow x + 2 = 0$ or $x - 4 = 0$

$\Rightarrow x = -2$ or $x = 4$

Factorising Quadratics

A) **Rearrange the equation into the standard $ax^2 + bx + c$ form.**

B) **Write down the two brackets:**
   (x   )(x   )

C) **Find two numbers that multiply to give 'c' and add / subtract to give 'b' (ignoring signs).**

D) **Put the numbers in the brackets and choose their signs.**

Another Example...

EXAMPLE: Solve $x^2 + 4x - 21 = 0$ by factorising.

This equation is already in the standard format — you can write down the brackets straight away.

$x^2 + 4x - 21 = (x\ \ \ \ )(x\ \ \ \ )$

This is the value of 'b' you're after — 3 and 7 are the right numbers.

1 and 21 multiply to give 21 — and add / subtract to give 22 and 20.

3 and 7 multiply to give 21 — and add / subtract to give 10 and 4.

$x^2 + 4x - 21 = (x + 7)(x - 3)$

And solving the equation to find x gives... $\Rightarrow x = -7$ or $x = 3$

Scitardauq Gnisirotcaf — you should know it backwards...

Factorising quadratics — this is <u>very</u> basic stuff. You've really got to be comfortable with it. If you're even slightly rusty, you need to practise it until it's second nature. Remember why you're doing it — you don't factorise simply for the pleasure it gives you — it's so you can <u>solve</u> quadratic equations. Well, that's the theory anyway...

Factorising a Quadratic

It's not over yet...

Factorising a quadratic when $a \neq 1$

These can be a real pain. The basic method's the same as on the previous page — but it can be a bit more awkward.

EXAMPLE: Factorise $3x^2 + 4x - 15$

A   Write Down Two Brackets

As before, write down two brackets — but instead of just having x in each, you need two things that will multiply to give $3x^2$.

It's got to be $3x$ and x here.

$$3x^2 + 4x - 15 = (3x \quad)(x \quad)$$

B   The Fiddly Bit

You need to find two numbers that multiply together to make 15 — but which will give you 4x when you multiply them by x and 3x, and then add / subtract them.

$(3x \quad 1)(x \quad 15) \Rightarrow x$ and $45x$ which then add or subtract to give 46x and 44x.

$(3x \quad 15)(x \quad 1) \Rightarrow 15x$ and $3x$ which then add or subtract to give 18x and 12x.

$(3x \quad 3)(x \quad 5) \Rightarrow 3x$ and $15x$ which then add or subtract to give 18x and 12x.

$(3x \quad 5)(x \quad 3) \Rightarrow 5x$ and $9x$ which then add or subtract to give 14x and 4x.

This is the value you're after — so this is the right combination.

C   Add the Signs

You know the brackets must be like these... $\Rightarrow (3x \quad 5)(x \quad 3) = 3x^2 + 4x - 15$

'c' is negative — that means the signs in the brackets are different.

So all you have to do is put in the plus or minus signs.

You've only got two choices — if you're unsure, just multiply them out to see which one's right.

$$(3x + 5)(x - 3) = 3x^2 - 4x - 15$$

or...

$$(3x - 5)(x + 3) = 3x^2 + 4x - 15 \quad \Leftarrow \text{So it's this one.}$$

Sometimes it's best just to Cheat and use the Formula

Here's two final points to bear in mind:

1) It won't always factorise.

2) Sometimes factorising is so messy that it's easier to just use the quadratic formula...

So if the question doesn't tell you to factorise, don't assume it will factorise.
And if it's something like this thing below, don't bother trying to factorise it...

EXAMPLE: Solve $6x^2 + 87x - 144 = 0$

This will actually factorise, but there's 2 possible bracket forms to try.

$(6x \quad)(x \quad)$ or $(3x \quad)(2x \quad)$ And for each of these, there's 8 possible ways of making 144 to try.

And you can quote me on that...

"He who can properly do quadratic equations is considered a god."
Plato

"Quadratic equations are the music of reason."
James J Sylvester

Completing the Square

Completing the Square is a handy little trick that you should <u>definitely</u> know how to use.
It can be a bit fiddly — but it gives you <u>loads</u> of information about a quadratic really quickly.

Take any old quadratic and put it in a **Special Form**

Completing the square can be really confusing. For starters, what does "Completing the Square" <u>mean</u>?
<u>What</u> is the square? <u>Why</u> does it need completing? Well, there is <u>some</u> logic to it:

1) The <u>square</u> is something like this: $(x + \text{something})^2$ It's basically the factorised equation (with the factors both the same), but there's something missing...

2) ...So you need to '<u>complete</u>' it by adding a number to the square to make it equal to the original expression. $(x + \text{something})^2 + d$

You'll start with something like this...   ...sort the x-coefficients...   ...and you'll end up with something like this.

$$2x^2 + 8x - 5 \quad \Longrightarrow \quad 2(x+2)^2 + ? \quad \Longrightarrow \quad 2(x+2)^2 - 13$$

Lovely!

Make completing the square a bit **Easier**

There are only a few stages to completing the square — if you can't be bothered trying to understand it,
just <u>learn how to do it</u>. But I reckon it's worth spending a bit more time to get your head round it <u>properly</u>.

A

Take Out a Factor of 'a'

— take a factor of a out of the x^2 and x terms.

This '2' is an 'a'.

$$f(x) = 2x^2 + 3x - 5$$ This is in the form $ax^2 + bx + c$

$$f(x) = 2\left(x^2 + \tfrac{3}{2}x\right) - 5$$ Check that the bracket multiplies out to what you had before.

This is $\frac{b}{a}$

B

Rewrite the Bracket   — rewrite the bracket as one bracket squared.

The number in the brackets is <u>always</u> half the old number in front of the x. $\dfrac{b}{2a}$

$$f(x) = 2\left(x + \tfrac{3}{4}\right)^2 + d$$ d is a number you have to find to make the new form equal to the old one.

Don't forget the 'squared' sign.

C

Complete the Square   — find d.

To do this, <u>make the old and new equations equal each other</u>...

...and you can find d.

The x^2 and x bits are the same on both sides so they can disappear.

$$2\left(x + \tfrac{3}{4}\right)^2 + d = 2x^2 + 3x - 5$$

$$2x^2 + 3x + \tfrac{9}{8} + d = 2x^2 + 3x - 5$$

$$\tfrac{9}{8} + d = -5$$

$$\Rightarrow d = -\tfrac{49}{8}$$

Completing the Square

A) <u>THE BIT IN THE BRACKETS IS ALWAYS</u> — $a\left(x + \dfrac{b}{2a}\right)^2$

B) <u>CALL THE NUMBER AT THE END d</u> — $a\left(x + \dfrac{b}{2a}\right)^2 + d$

C) <u>MAKE THE TWO FORMS EQUAL</u> — $ax^2 + bx + c = a\left(x + \dfrac{b}{2a}\right)^2 + d$

D

So the Answer is: $f(x) = 2x^2 + 3x - 5 = 2\left(x + \tfrac{3}{4}\right)^2 - \dfrac{49}{8}$

Complete your square — it'd be root not to...

Remember — you're basically trying to write the expression as one bracket squared, but it doesn't quite work. So you have
to add a number (d) to make it work. It's a bit confusing at first, but once you've learnt it you won't forget it in a hurry.

Completing the Square

Once you've completed the square, you can very quickly say <u>loads</u> about a quadratic function.
And it all relies on the fact that a squared number can <u>never</u> be less than zero... <u>ever</u>.

Completing the square can sometimes be Useful

This is a quadratic written as a completed square. As it's a quadratic
function and the coefficient of x^2 is positive, it's a u-shaped graph.

This is a square — it can never be negative. The smallest it can be is 0.

$$f(x) = 3x^2 - 6x - 7 = 3(x-1)^2 - 10$$

A

Find the Minimum — make the bit in the brackets equal to zero.

When the squared bit is zero, f(x)
reaches its minimum value.
This means the graph reaches its
lowest point.

$$f(x) = 3(x-1)^2 - 10$$

This number here is the minimum.

$$f(1) = 3(1-1)^2 - 10$$

f(1) means using x=1 in the function

$$f(1) = 3(0)^2 - 10 = -10$$

So the minimum is -10, when x=1

B

Where Does f(x) Cross the x-axis? — i.e. find d.

Make the completed square
function equal zero.

$$3(x-1)^2 - 10 = 0$$

Solve it to find where f(x)
crosses the x-axis.

$$\Rightarrow (x-1)^2 = \frac{10}{3}$$

da-de-dah ... rearranging again.

$$\Rightarrow x - 1 = \pm\sqrt{\frac{10}{3}}$$

$$\Rightarrow x = 1 \pm \sqrt{\frac{10}{3}}$$

So f(x) crosses the x-axis when...

$$x = 1 + \sqrt{\frac{10}{3}} \text{ or } x = 1 - \sqrt{\frac{10}{3}}$$

These notes are all about graphs with <u>positive</u> coefficients in front of the x^2. But if the coefficient is negative, then the graph is flipped <u>upside-down</u> (n-shaped, not u-shaped).

With this information, you can
easily sketch the graph...

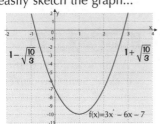

$1 - \sqrt{\frac{10}{3}}$ $1 + \sqrt{\frac{10}{3}}$

$f(x)=3x^2 - 6x - 7$

Some functions don't have Real Roots

By completing the square, you can also quickly tell if the graph of a quadratic function ever crosses the x-axis.
It'll only cross the x-axis if the function changes sign (i.e. goes from positive to negative or vice versa).
Take this function...

Find the Roots

$$f(x) = x^2 + 4x + 7$$

This number's positive.

$$f(x) = (x+2)^2 + 3$$

The smallest this bit can be is zero (at x = −2).

$(x + 2)^2$ is never less than zero so f(x) is never less than three.

This means that:

a) f(x) can <u>never</u> be negative.
b) The graph of f(x) <u>never</u> crosses the x-axis.

If the coefficient of x^2 is negative, you can do the same sort of thing to check whether f(x) ever becomes positive.

Don't forget — two wrongs don't make a root...

You'll be pleased to know that that's the end of me trying to tell you how to do something you probably really don't
want to do. Now you can push it to one side and run off to roll around in a bed of nettles... much more fun.

The Quadratic Formula

Unlike factorising, the quadratic formula always works... no ifs, no buts, no butts, no nothing...

The **Quadratic Formula** — *a reason to be cheerful, but careful...*

If you want to solve a quadratic equation $ax^2 + bx + c = 0$,
then the answers are given by this formula:

$$x = \frac{-b \pm \sqrt{b^2 - 4ac}}{2a}$$

This formula will NOT be in your formula book come the exam. You need to learn it.

The formula's a godsend — but use the power wisely...

If any of the coefficients (i.e. if a, b or c) in your quadratic equation are negative — be <u>especially</u> careful.

Always take things nice and <u>slowly</u> — don't try to rush it.

It's a good idea to write down what a, b and c are <u>before</u> you start plugging them into the formula.

There are a couple of minus signs in the formula — which can catch you out if you're not paying <u>attention</u>.

I shall teach you the ways of the **Formula**

EXAMPLE: Solve the quadratic equation $3x^2 - 4x = 8$, leaving your answers in surd form.

The mention of <u>surds</u> is a <u>big</u> clue that you should use the formula.

Rearrange the Equation

Get the equation in the standard $ax^2 + bx + c = 0$ form.

$3x^2 - 4x = 8$

$3x^2 - 4x - 8 = 0$

Find a, b and c

Write down the coefficients a, b and c — making sure you don't forget minus signs.

$3x^2 - 4x - 8 = 0$

$a = 3 \qquad b = -4 \qquad c = -8$

Stick Them in the Formula

Very carefully, plug these numbers into the formula.
It's best to write down each stage as you do it.

$$x = \frac{-b \pm \sqrt{b^2 - 4ac}}{2a}$$

$$= \frac{-(-4) \pm \sqrt{(-4)^2 - 4 \times 3 \times (-8)}}{2 \times 3}$$

$$= \frac{4 \pm \sqrt{16 + 96}}{6}$$

$$= \frac{4 \pm \sqrt{112}}{6}$$

The $\pm$ sign means that we have two different expressions for x — which you get by replacing the $\pm$ with + and –.

$$= \frac{4 \pm \sqrt{16 \times 7}}{6} = \frac{4 \pm 4\sqrt{7}}{6} = \frac{2}{3} \pm \frac{2}{3}\sqrt{7}$$

$$x = \frac{2}{3} + \frac{2}{3}\sqrt{7} \text{ or } x = \frac{2}{3} - \frac{2}{3}\sqrt{7}$$

Using this magic formula, I shall take over the world... ha ha ha...

Okay, maybe it's not <u>quite</u> that good... but it's really important. So learn it properly — which means spending enough time until you can just say it out loud the whole way through, with no hesitations. Or perhaps you could try singing it as loud as you can to the tune of your favourite cheesy song. Sha-la-la-la-la-la-la-ha... La-di-da... Sha-la-la-la-la-la-la-ha... La-di-da... Sha-la-la-la-la-la-la-ha...

The Quadratic Formula

By using part of the quadratic formula, you can quickly tell if a quadratic equation has two solutions, one solution, or no solutions at all. Tell me more, I hear you cry...

How Many Roots? Check the b² – 4ac bit...

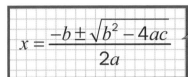

$$x = \frac{-b \pm \sqrt{b^2 - 4ac}}{2a}$$

When you try to find the roots of a quadratic function, this bit in the square-root sign ($b^2 - 4ac$) can be positive, zero, or negative. It's <u>this</u> that tells you if a quadratic function has two roots, one root, or no roots.

The $b^2 - 4ac$ bit is called the <u>discriminant</u>.

<u>Because</u> — if the discriminant is positive, the formula will give you two different values — when you add or subtract the $\sqrt{b^2 - 4ac}$ bit.

<u>But</u> if it's zero, you'll only get one value, since adding or subtracting zero doesn't make any difference.

<u>And</u> if it's negative, you don't get any (real) values because you can't take the square root of a negative number.

Well, not in Core 1. In later modules, you can actually take the square root of negative numbers and get 'imaginary' numbers. That's why we say no 'real' roots — because there are 'imaginary' roots!

It's good to be able to picture what this means:

A root is just when y = 0, so it's where the graph touches or crosses the x-axis.

$b^2 - 4ac > 0$	$b^2 - 4ac = 0$	$b^2 - 4ac < 0$
Two roots	One root	No roots

So the graph crosses the x-axis twice and these are the roots:

The graph just touches the x-axis from above (or from below if the x² coefficient is negative).

The graph doesn't touch the x-axis at all.

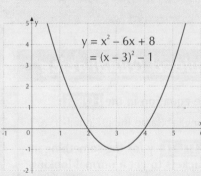

$y = x^2 - 6x + 8$
$= (x - 3)^2 - 1$

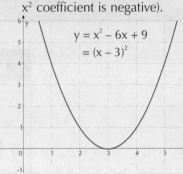

$y = x^2 - 6x + 9$
$= (x - 3)^2$

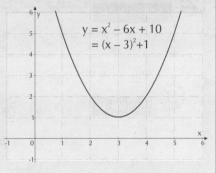

$y = x^2 - 6x + 10$
$= (x - 3)^2 + 1$

EXAMPLE: Find the range of values of k for which: a) f(x)=0 has 2 distinct roots, b) f(x)=0 has 1 root, c) f(x) has no real roots, where $f(x) = 3x^2 + 2x + k$.

First of all, work out what the discriminant is:
$$b^2 - 4ac = 2^2 - 4 \times 3 \times k$$
$$= 4 - 12k$$

These calculations are exactly the same. You don't need to do them if you've done a) because the only difference is the equality symbol.

a) <u>Two distinct roots</u> means:
$$b^2 - 4ac > 0 \Rightarrow 4 - 12k > 0$$
$$\Rightarrow 4 > 12k$$
$$\Rightarrow k < \tfrac{1}{3}$$

b) <u>One root</u> means:
$$b^2 - 4ac = 0 \Rightarrow 4 - 12k = 0$$
$$\Rightarrow 4 = 12k$$
$$\Rightarrow k = \tfrac{1}{3}$$

c) <u>No roots</u> means:
$$b^2 - 4ac < 0 \Rightarrow 4 - 12k < 0$$
$$\Rightarrow 4 < 12k$$
$$\Rightarrow k > \tfrac{1}{3}$$

---

ha ha ha ha haaaaaa... ha ha ha... ha ha ha ... ha ha ha........

So for questions about "how many roots", think discriminant — i.e. $b^2 - 4ac$. And don't get the inequality signs (> and <) the wrong way round. It's obvious, if you think about it.

'Almost' Quadratic Equations

Sometimes you'll be asked to solve equations that look really difficult, like the ones on this page. But with a bit of rearrangement and fiddling you can get them to look just like an ordinary quadratic you can solve.

Some Nasty-looking equations are just Quadratics

$$x^4 + 3x^2 + 6 = 0$$

Arrrgh. How on earth are you supposed to solve something like that? Well the answer is... with great difficulty — that's if you don't spot that you can turn it into quadratic form like this:

$$(x^2)^2 + 3(x^2) + 6 = 0$$

It still looks weird. But, if those x²'s were y's:

$$y^2 + 3y + 6 = 0$$

Now it's a just a simple quadratic that you could solve in your sleep — or the exam, which would probably be more useful.

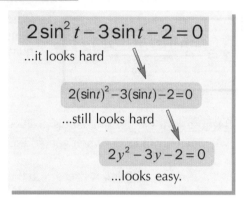

$$2\sin^2 t - 3\sin t - 2 = 0$$

...it looks hard

$$2(\sin t)^2 - 3(\sin t) - 2 = 0$$

...still looks hard

$$2y^2 - 3y - 2 = 0$$

...looks easy.

Just make a Substitution to Simplify

Example: $2x^6 - 11x^3 + 5 = 0$

1 **Spot That It's a Quadratic**

Put it in the form: a(something)² + b(same thing) + (number) = 0.

$$2(x^3)^2 - 11(x^3) + 5 = 0$$

Now substitute x³ for y to make it like a normal quadratic.

2 **Substitute**

let $x^3 = y \Rightarrow$ $2y^2 - 11y + 5 = 0$

And solve this quadratic to find the values of y.

3 **Solve it**

$$2y^2 - 11y + 5 = 0$$

$$(2y - 1)(y - 5) = 0$$

$$y = \tfrac{1}{2}, \text{ or } 5$$

Now you've got the values of y, you can get the values of x.

4

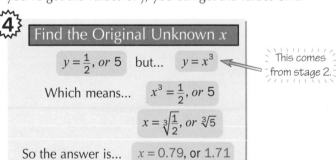

Find the Original Unknown x

$y = \tfrac{1}{2}, \text{ or } 5$ but... $y = x^3$ ← This comes from stage 2.

Which means... $x^3 = \tfrac{1}{2}, \text{ or } 5$

$$x = \sqrt[3]{\tfrac{1}{2}}, \text{ or } \sqrt[3]{5}$$

So the answer is... $x = 0.79$, or 1.71

Disguised Quadratics

1) **Put the equation in the FORM :**

   a(something)²+b(same thing)+(number)=0

2) **SUBSTITUTE y for the something in the brackets to get a normal-looking quadratic.**

3) **SOLVE the quadratic in the usual way — i.e. by factorising or using the quadratic formula.**

4) **Stick your answers in the substitution equation to get the values for the ORIGINAL unknown.**

---

almost quadratics — almost worthwhile, almost interesting, almost...

Quadratics with delusions of grandeur. Whatever next. Anyway, I haven't really got anything to add to what's already on this page. But it's been one long, mean section. "Nuff about quadratics!" I hear you cry. Well it's all over now, so you can relax...
ahhhhhh   ....just relax....   ....mmm.....   ....nice......   ......ah.....   Hang on. You've still got the revision questions to do.

Section Two Revision Questions

Mmmm, well, quadratic equations — not exactly designed to make you fall out of your chair through laughing so hard, are they? But (and that's a huge 'but') they'll get you plenty of marks come that fine morning when you march confidently into the exam hall — if you know what you're doing. And what better way to make sure you know what you're doing than to practise. So here we go then, on the thrill-seekers' ride of a lifetime — the CGP quadratic equation revision section...

1) Factorise the following expressions. While you're doing this, sing a jolly song to show how much you enjoy it.

    a) $x^2 + 2x + 1$,   b) $x^2 - 13x + 30$,   c) $x^2 - 4$,   d) $3 + 2x - x^2$

    e) $2x^2 - 7x - 4$,   f) $5x^2 + 7x - 6$.

2) Solve the following equations. And sing verse two of your jolly song.

    a) $x^2 - 3x + 2 = 0$,   b) $x^2 + x - 12 = 0$,   c) $2 + x - x^2 = 0$,   d) $x^2 + x - 16 = x$

    e) $3x^2 - 15x - 14 = 4x$,   f) $4x^2 - 1 = 0$,   g) $6x^2 - 11x + 9 = 2x^2 - x + 3$.

3) Rewrite these quadratics by completing the square. Then state their maximum or minimum value and the value of x where this occurs. Also, say which ones cross the x-axis — just for a laugh, like.

    a) $x^2 - 4x - 3$,   b) $3 - 3x - x^2$,   c) $2x^2 - 4x + 11$,   d) $4x^2 - 28x + 48$.

4) How many roots do these quadratics have? Sketch their graphs.

    a) $x^2 - 2x - 3 = 0$,   b) $x^2 - 6x + 9 = 0$,   c) $2x^2 + 4x + 3 = 0$.

5) Solve these quadratic equations, leaving your answers in surd form.

    a) $3x^2 - 7x + 3 = 0$,   b) $2x^2 - 6x - 2 = 0$,   c) $x^2 + 4x + 6 = 12$.

6) If the quadratic equation $x^2 + kx + 4 = 0$ has two real roots, what are the possible values of k?

7) Find all the solutions to the following equations. (Quite tricky, these — especially that last one.)

    a) $2\sin^2 x - \sin x - 1 = 0$, for $-90° \le x \le 90°$,   b) $x^4 - 17x^2 + 16 = 0$,   c) $x^{\frac{4}{3}} - 5x^{\frac{2}{3}} + 4 = 0$.

    OK, I think that's enough. Go and make yourself a cup of tea. Treat yourself to a chocolate biscuit.

Here is a new way to enjoy Pelican biscuits:

Bite a small piece off two opposite corners of a Pelican.
Immerse one corner in coffee, and suck coffee up through the Pelican,
like a straw. You will need to suck quite hard to start with.
After a few seconds you will notice the biscuit part of the Pelican start to
lose structural integrity. At this point, cram it into your mouth, where it
will collapse into a mass of hot molten chocolate, biscuit and coffee.

Mmmm.

Linear Inequalities

Solving <u>inequalities</u> is very similar to solving equations. You've just got to be really careful that you keep the inequality sign pointing the <u>right</u> way.

> Find the ranges of x that satisfy these inequalities:
>
> (i) $x - 3 < -1 + 2x$ (ii) $8x + 2 \geq 2x + 17$ (iii) $4 - 3x \leq 16$ (iv) $36x < 6x^2$

Sometimes the inequality sign **Changes Direction**

Like I said, these are pretty similar to solving equations — because whatever you do to one side, you have to do to the other. But multiplying or dividing by <u>negative</u> numbers <u>changes</u> the direction of the inequality sign.

Adding or **Subtracting** doesn't change the direction of the inequality sign

EXAMPLE: If you <u>add</u> or <u>subtract</u> something from both sides of an inequality, the inequality sign <u>doesn't</u> change direction.

Adding 1 to both sides leaves the inequality sign pointing in the same direction.

Subtracting x from both sides doesn't affect the inequality.

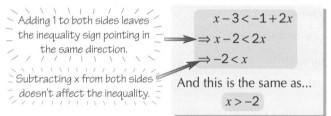

$$x - 3 < -1 + 2x$$
$$\Rightarrow x - 2 < 2x$$
$$\Rightarrow -2 < x$$

And this is the same as...

$$x > -2$$

Multiplying or **Dividing** by something **Positive** doesn't affect the inequality sign

EXAMPLE: Multiplying or dividing both sides of an inequality by a <u>positive</u> number <u>doesn't</u> affect the direction of the inequality sign.

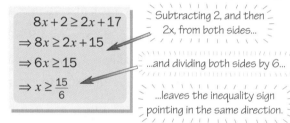

$$8x + 2 \geq 2x + 17$$
$$\Rightarrow 8x \geq 2x + 15$$
$$\Rightarrow 6x \geq 15$$
$$\Rightarrow x \geq \frac{15}{6}$$

Subtracting 2, and then 2x, from both sides...

...and dividing both sides by 6...

...leaves the inequality sign pointing in the same direction.

But **Change** the inequality if you **Multiply** or **Divide** by something **Negative**

But multiplying or dividing both sides of an inequality by a <u>negative</u> number <u>changes</u> the direction of the inequality.

EXAMPLE:

$$4 - 3x \leq 16$$
$$\Rightarrow -3x \leq 12$$
$$\Rightarrow x \geq -4$$

Subtract 4 from both sides.

Then divide both sides by –3 — but <u>change</u> the direction of the inequality.

> The <u>reason</u> for the sign changing direction is because it's just the same as swapping everything from one side to the other:
>
> $-3x \leq 12$ $\Rightarrow -12 \leq 3x$ $\Rightarrow x \geq -4$

Don't divide both sides by **Variables** — like x and y

You've got to be really careful when you divide by things that <u>might</u> be negative — well basically, don't do it.

EXAMPLE: $36x < 6x^2$

Start by dividing by 6.

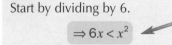

$$\Rightarrow 6x < x^2$$

Dividing by 6 is okay because 6 is definitely positive.

It's tempting to divide both sides by x now — but x could be negative (or zero).

So instead... $0 < x^2 - 6x$

It's much safer to take 6x from both sides and solve this...

Which is... $x^2 - 6x > 0$

Two types of inequality sign

There are two kinds of inequality sign:

    Type 1:   < — less than
                > — greater than

    Type 2:   ≤ — less than or equal to
                ≥ — greater than or equal to

Whatever type the question uses — use the same kind all the way through your answer.

See the next page for more on solving quadratic inequalities.

So no one knows we've arrived safely — splendid...

So just remember — inequalities are just like normal equations except that you have to reverse the sign when multiplying or dividing by a negative number. And <u>don't</u> divide both sides by variables. (You should know not to do this with normal equations anyway because the variable could be <u>zero</u>.) OK — lecture's over.

Quadratic Inequalities

With quadratic inequalities, you're best off drawing the <u>graph</u> and taking it from there.

Draw a **Graph** to solve a **Quadratic** inequality

<u>Example:</u> Find the ranges of x which satisfy these inequalities:

① $-x^2 + 2x + 4 \geq 1$ ② $2x^2 - x - 3 > 0$

First rewrite the inequality with <u>zero</u> on one side.

$-x^2 + 2x + 3 \geq 0$

Then <u>draw</u> the graph of $y = -x^2 + 2x + 3$:

So find where it crosses the x-axis (i.e. where y=0):

$-x^2 + 2x + 3 = 0 \Rightarrow x^2 - 2x - 3 = 0$
$\Rightarrow (x+1)(x-3) = 0$
$\Rightarrow x = -1 \text{ or } x = 3$

And the coefficient of x^2 is negative, so the graph is n-shaped. So it looks like this:

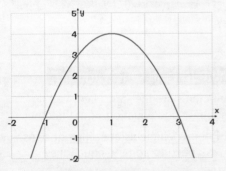

You're interested in when this is <u>positive or zero</u>, i.e. when it's above the x-axis.

From the graph, this is when x is <u>between –1 and 3</u> (including those points). So your answer is...

$-x^2 + 2x + 4 \geq 1$ when $-1 \leq x \leq 3$.

This one already has zero on one side, so <u>draw</u> the graph of $y = 2x^2 - x - 3$.

Find where it crosses the x-axis:

$2x^2 - x - 3 = 0$
$\Rightarrow (2x - 3)(x + 1)$
$\Rightarrow x = \frac{3}{2} \text{ or } x = -1$

Factorise it to find the roots.

And the coefficient of x^2 is positive, so the graph is u-shaped. And looks like this:

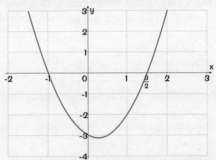

You need to say when this is <u>positive</u>. Looking at the graph, there are two parts of the x-axis where this is true — when x is <u>less than –1</u> and when x is <u>greater than 3/2</u>. So your answer is:

$2x^2 - x - 3 > 0$ when $x < -1$ or $x > \frac{3}{2}$.

<u>Example (revisited):</u>

On the last page you had to solve $36x < 6x^2$.

$36x < 6x^2$

equation 1 $\Longrightarrow$ $\Rightarrow 6x < x^2$
$\Rightarrow 0 < x^2 - 6x$

So draw the graph of

$y = x^2 - 6x = x(x - 6)$

And this is <u>positive</u> when $x < 0$ or $x > 6$.

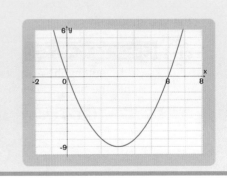

If you divide by x in equation 1, you'd only get half the solution — you'd miss the $x < 0$ part.

That's nonsense — I can see perfectly...

Call me sad, but I reckon these questions are pretty cool. They look a lot more difficult than they actually are and you get to draw a picture. Wow! When you do the graph, the important thing is to find where it crosses the x-axis (you don't need to know where it crosses the y-axis) and make sure you draw it the right way up. Then you just need to decide which bit of the graph you want. It'll either be the range(s) of x where the graph is below the x-axis or the range(s) where it's above. And this depends on the inequality sign.

Simultaneous Equations

Solving simultaneous equations means finding the answers to two equations <u>at the same time</u> — i.e. finding values for x and y for which both equations are true. And it's one of those things that you'll have to do <u>again and again</u> — so it's definitely worth practising them until you feel <u>really confident</u>.

① $3x + 5y = -4$
② $-2x + 3y = 9$

This is how simultaneous equations are usually shown. It's a good idea to label them as equation ① and equation ② — so you know which one you're working with.

But they'll look different sometimes, maybe like this. $\longrightarrow$ $4 + 5y = -3x$ $\quad$ rearrange as $3x + 5y = -4$
Make sure you rearrange them as 'ax + by = c'. $\quad -2x = 9 - 3y$ $\xrightarrow{ax + by = c}$ $-2x + 3y = 9$

Solving them by *Elimination*

Elimination is a lovely method. It's really quick when you get the hang of it
— you'll be doing virtually all of it in your head.

EXAMPLE:

① $\quad 3x + 5y = -4$
② $\quad -2x + 3y = 9$

To get the x's to match, you need to multiply the first equation by 2 and the second by 3:

① ×2 $\quad 6x + 10y = -8$
② ×3 $\quad -6x + 9y = 27$

Add the equations together to eliminate the x's.

① + ② $\quad 19y = 19$
$\quad\quad\quad y = 1$

So y is 1. Now stick that value for y into one of the equations to find x:

$y = 1$ in ① $\Rightarrow 3x + 5 = -4$
$\quad\quad\quad 3x = -9$
$\quad\quad\quad x = -3$

So the solution is x = –3, y = 1.

A | Match the Coefficients

Multiply the equations by numbers that will make either the x's or the y's match in the two equations. (Ignoring minus signs.)

Go for the lowest common multiple (LCM).
e.g. LCM of 2 and 3 is 6.

B | Eliminate to Find One Variable

If the coefficients are the <u>same</u> sign, you'll need to <u>subtract</u> one equation from the other.

If the coefficients are <u>different</u> signs, you need to <u>add</u> the equations.

C | Find the Variable You Eliminated

When you've found one variable, put its value into one of the original equations so you can find the other variable.

But you should always...

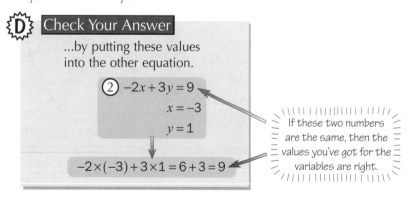

D | Check Your Answer

...by putting these values into the other equation.

② $-2x + 3y = 9$
$\quad x = -3$
$\quad y = 1$

$-2 \times (-3) + 3 \times 1 = 6 + 3 = 9$

If these two numbers are the same, then the values you've got for the variables are right.

Elimination Method

1) <u>Match the coefficients</u>

2) <u>Eliminate and then solve for one variable</u>

3) <u>Find the other variable (that you eliminated)</u>

4) <u>Check your answer</u>

Eliminate your social life — do AS-level maths

This is a fairly basic method that won't be new to you. So make sure you know it. The only possibly tricky bit is matching the coefficients — work out the lowest common multiple of the coefficients of x, say, then multiply the equations to get this number in front of each x.

Simultaneous Equations with Quadratics

Elimination is great for simple equations. But it won't always work. Sometimes one of the equations has not just x's and y's in it — but bits with x^2 and y^2 as well. When this happens, you can <u>only</u> use the <u>substitution</u> method.

Use Substitution if one equation is **Quadratic**

EXAMPLE: $-x + 2y = 5$ ——Ⓛ ◄—— The <u>linear</u> equation — with only x's and y's in.

$x^2 + y^2 = 25$ ——Ⓠ ◄—— The <u>quadratic</u> equation — with some x^2 and y^2 bits in.

Rearrange the <u>linear equation</u> so that either x or y is on its own on one side of the equals sign.

Ⓛ $-x + 2y = 5$
$\Rightarrow x = 2y - 5$

Substitute this expression into the <u>quadratic equation</u>...

Sub into Ⓠ: $x^2 + y^2 = 25$
$\Rightarrow (2y - 5)^2 + y^2 = 25$

...and then rearrange this into the form $ax^2 + bx + c = 0$, so you can solve it — either by <u>factorising</u> or using the <u>quadratic formula</u>.

$\Rightarrow (4y^2 - 20y + 25) + y^2 = 25$
$\Rightarrow 5y^2 - 20y = 0$
$\Rightarrow 5y(y - 4) = 0$
$\Rightarrow y = 0$ or $y = 4$

One Quadratic and One Linear Eqn

1) **Isolate variable in linear equation**
Rearrange the linear equation to get either x or y on its own.

2) **Substitute into quadratic equation**
— to get a quadratic equation in just one variable.

3) **Solve to get values for one variable**
— either by factorising or using the quadratic formula.

4) **Stick these values in the linear equation**
— to find corresponding values for the other variable.

Finally put both these values back into the <u>linear equation</u> to find corresponding values for x:

When y = 0: $-x + 2y = 5$ Ⓛ
$\Rightarrow x = -5$

When y = 4: $-x + 2y = 5$ Ⓛ
$\Rightarrow -x + 8 = 5$
$\Rightarrow x = 3$

So the solutions to the simultaneous equations are: x = –5, y = 0 and x = 3, y = 4.

As usual, <u>check your answers</u> by putting these values back into the original equations.

Check Your Answers

x = -5, y = 0: $-(-5) + 2 \times 0 = 5$ ✓
$(-5)^2 + 0^2 = 25$ ✓

x = 3, y = 4: $-(3) + 2 \times 4 = 5$ ✓
$3^2 + 4^2 = 25$ ✓

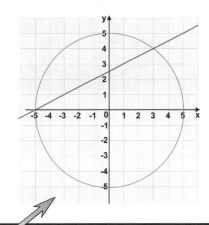

$y = x^2$ — a match-winning substitution...

The quadratic equation above is actually a <u>circle</u> about the origin with radius 5. (Don't worry, the pages about circles come later in the book — see Section 4 for more info). The linear equation is just a standard straight line. So what you're actually finding here are the two points where the line passes through the circle. And these turn out to be (–5,0) and (3,4). See the graph. (I thought you might appreciate seeing a graph that wasn't a line or a parabola for a change.)

Cows

The stuff on this page isn't strictly on the syllabus. But I've included it anyway because I reckon it's really important stuff that you ought to know.

There are loads of Different Types of Cows

Dairy Cattle

Every day a dairy cow can produce up to 128 pints of milk — which can be used to make 14 lbs of cheese, 5 gallons of ice cream, or 6 lbs butter.

The Jersey
The Jersey is a small breed best suited to pastures in high rainfall areas. It is kept for its creamy milk.

Advantages
1) Can produce creamy milk until old age.
2) Milk is the highest in fat of any dairy breed (5.2%).
3) Fairly docile, although bulls can't be trusted.

Disadvantages
1) Produces less milk than most other breeds.

The Holstein-Friesian
This breed can be found in many areas. It is kept mainly for milk.

Advantages
1) Produce more milk than any breed.
2) The breed is large, so bulls can be sold for beef.

Disadvantages
1) Milk is low in fat (3.5%).

Beef Cattle

Cows are sedentary animals who spend up to 8 hours a day chewing the cud while standing still or lying down to rest after grazing. Getting fat for people to eat.

The Angus
The Angus is best suited to areas where there is moderately high rainfall.

Advantages
1) Early maturing.
2) High ratio of meat to body weight.
3) Forages well.
4) Adaptable.

The Hereford
The Hereford matures fairly early, but later than most shorthorn breeds. All Herefords have white faces, and if a Hereford is crossbred with any other breed of cow, all the offspring will have white or partially white faces.

Advantages
1) Hardy.
2) Adaptable to different feeds.

Disadvantages
1) Susceptible to eye diseases.

This is *really* important — try not to forget it.

Milk comes from Cows

Milk is an emulsion of butterfat suspended in a solution of water (roughly 80%), lactose, proteins and salts. Cow's milk has a specific gravity around 1.03.

It's pasteurised by heating it to 63°C for 30 minutes. It's then rapidly cooled and stored below 10°C.

Louis Pasteur began his experiments into 'pasteurisation' in 1856. By 1946, the vacuum pasteurisation method had been perfected, and in 1948, UHT (ultra heat-treated) pasteurisation was introduced.

$$cow + grass = fat\ cow$$
$$fat\ cow + milking\ machine \Rightarrow milk$$

You will often see cows with pieces of grass sticking out of their mouths.

SOME IMPORTANT FACTS TO REMEMBER:
• A newborn calf can walk on its own an hour after birth
• A cow's teeth are only on the bottom of her mouth
• While some cows can live up to 40 years, they generally don't live beyond 20.

Cows on the Internet

For more information on cows, try these websites:

www.allcows.com (including Cow of the Month)
www.crazyforcows.com (with cow e-postcards)
www.moomilk.com (includes a 'What's the cow thinking?' contest.)
http://www.geocities.com/Hollywood/9317/meowcow.html
(for cow-tipping on the Internet)

The Cow
The cow is of the bovine ilk;
One end is moo,
the other, milk.

— Ogden Nash

Famous Cows and Cow Songs

Famous Cows
1) Ermintrude from the Magic Roundabout.
2) Graham Heifer — the Boddingtons cow.
3) Other TV commercial cows — Anchor, Dairylea
4) The cow that jumped over the moon.
5) Greek Mythology was full of gods turning themselves and their girlfriends into cattle.

Cows in Pop Music
1) Size of a Cow — the Wonder Stuff
2) Saturday Night at the Moo-vies — The Drifters
3) What can I do to make you milk me? — The Cows
4) One to an-udder — the Charlatans
5) Milk me baby, one more time — Britney Spears

Where's me Jersey — I'm Friesian...

Cow-milking — an underrated skill, in my opinion. As Shakespeare once wrote, 'Those who can milk cows are likely to get pretty good grades in maths exams, no word of a lie'. Well, he probably would've written something like that if he was into cows. And he would've written it because cows are helpful when you're trying to work out what a question's all about — and once you know that, you can decide the best way forward. And if you don't believe me, remember the saying of the ancient Roman Emperor Julius Caesar, 'If in doubt, draw a cow'.

Geometric Interpretation

When you have to interpret something <u>geometrically</u> — you have to draw a picture and 'say what you see'.

Two Solutions — Two points of Intersection

Example:
$$y = x^2 - 4x + 5 \quad ①$$
$$y = 2x - 3 \quad ②$$

Solution:

Substitute expression for y from ② into ①:
$$2x - 3 = x^2 - 4x + 5$$

Rearrange and solve:
$$x^2 - 6x + 8 = 0$$
$$(x - 2)(x - 4) = 0$$
$$x = 2 \text{ or } x = 4$$

In ② gives:
$$x = 2 \Rightarrow y = 2 \times 2 - 3 = 1$$
$$x = 4 \Rightarrow y = 2 \times 4 - 3 = 5$$

There's 2 pairs of solutions: x=2, y=1 and x=4, y=5

Geometric Interpretation:

So from solving the simultaneous equations, you know that the graphs meet in <u>two places</u> — the points (2,1) and (4,5).

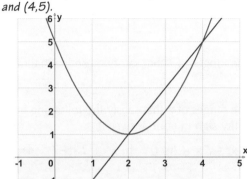

One Solution — One point of Intersection

Example:
$$y = x^2 - 4x + 5 \quad ①$$
$$y = 2x - 4 \quad ②$$

Solution:

Substitute ② in ①:
$$2x - 4 = x^2 - 4x + 5$$

Rearrange and solve:
$$x^2 - 6x + 9 = 0$$
$$(x - 3)^2 = 0$$
$$x = 3$$

Double root i.e. you only get 1 solution from the quadratic.

In Equation ② gives:
$$y = 2 \times 3 - 4$$
$$y = 2$$

There's 1 solution: x=3, y=2

Geometric Interpretation:

Since the equations have only one solution, the two graphs only meet at one point — (3,2). The straight line is a <u>tangent</u> to the curve.

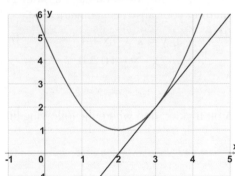

No Solutions means the Graphs Never Meet

Example:
$$y = x^2 - 4x + 5 \quad ①$$
$$y = 2x - 5 \quad ②$$

Solution:

Substitute ② in ①:
$$2x - 5 = x^2 - 4x + 5$$

Rearrange and try to solve with the quadratic formula:
$$x^2 - 6x + 10 = 0$$
$$b^2 - 4ac = (-6)^2 - 4.10$$
$$= 36 - 40 = -4$$

$b^2 - 4ac < 0$, so the quadratic has no roots.

So the simultaneous equations have no solutions.

Geometric Interpretation:

The equations have no solutions — the graphs never meet.

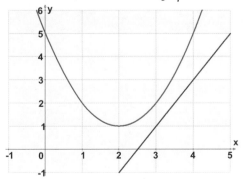

Geometric Interpretation? Frankly my dear, I don't give a damn...

There's some lovely practice Simultaneous Equations questions on the next page.

Section Three Revision Questions

What's that I hear you cry? You want revision questions — and lots of them. Well, it just so happens I've got a few here. Lots of questions on the different kinds of inequalities you need to know about, plus simultaneous equations. Now, as far as quadratic inequalities go, my advice is, 'if you're not sure, draw a picture — even if it's not accurate'. And as for simultaneous equations — well, just don't rush them — or you'll spend twice as long looking for milkshakes mistakes as it took you to do the question in the first place. That's it, the advice is over. So on with the questions...

1) Solve a) $7x - 4 > 2x - 42$, b) $12y - 3 \le 4y + 4$, c) $9y - 4 \ge 17y + 2$.

2) Find the ranges of x that satisfy these inequalities: i) $x + 6 < 5x - 4$ ii) $4x - 2 > x - 14$ iii) $7 - x \le 4 - 2x$

3) Find the ranges of x that satisfy the following inequalities. (And watch that you use the right kind of inequality sign in your answers.)

a) $3x^2 - 5x - 2 \le 0$, b) $x^2 + 2x + 7 > 4x + 9$, c) $3x^2 + 7x + 4 \ge 2(x^2 + x - 1)$.

4) Find the ranges of x that satisfy these jokers: i) $x^2 + 3x - 1 \ge x + 2$ ii) $2x^2 > x + 1$ iii) $3x^2 - 12 < x^2 - 2x$

5) Solve these sets of simultaneous equations.

a) $3x - 4y = 7$ and $-2x + 7y = -22$ b) $2x - 3y = \frac{11}{12}$ and $x + y = -\frac{7}{12}$

6) Find where possible (and that's a bit of a clue) the solutions to these sets of simultaneous equations. Interpret your answers geometrically.

a) $y = x^2 - 7x + 4$
$2x - y - 10 = 0$

b) $y = 30 - 6x + 2x^2$
$y = 2(x + 11)$

c) $x^2 + 2y^2 - 3 = 0$
$y = 2x + 4$

7) A bit trickier: find where the following lines meet:
a) $y = 3x - 4$ and $y = 7x - 5$,
b) $y = 13 - 2x$ and $7x - y - 23 = 0$,
c) $2x - 3y + 4 = 0$ and $x - 2y + 1 = 0$.

SECTION THREE — SIMULTANEOUS EQUATIONS & INEQUALITIES

Coordinate Geometry

Welcome to geometry club... nice — today I shall be mostly talking about straight lines...

Finding the equation of a line *Through Two Points*

If you get through your exam without having to find the equation of a line through two points, I'm a Dutchman.

EXAMPLE: Find the equation of the line that passes through the points (–3, 10) and (1, 4), and write it in the forms:

$$y - y_1 = m(x - x_1)$$

$$y = mx + c$$

$$ax + by + c = 0$$

— where a, b and c are <u>integers</u>.

You might be asked to write the equation of a line in <u>any</u> of these forms — but they're all similar.
Basically, if you find an equation in one form — you can easily <u>convert</u> it into either of the others.

The **Easiest** to find is $y - y_1 = m(x - x_1)$...

Point 1 is (–3, 10) and Point 2 is (1, 4)

Label the Points Label Point 1 as (x_1, y_1) and Point 2 as (x_2, y_2).

$$\text{Point 1} - (x_1, y_1) = (-3, 10)$$

$$\text{Point 2} - (x_2, y_2) = (1, 4)$$

It doesn't matter which way round you label them.

Find the Gradient Find the <u>gradient</u> of the line m — this is $m = \dfrac{y_2 - y_1}{x_2 - x_1}$.

$$m = \frac{4-10}{1-(-3)} = \frac{-6}{4} = -\frac{3}{2}$$

Write Down the Equation <u>Write down</u> the equation of the line, using the coordinates x_1 and y_1 — this is just $y - y_1 = m(x - x_1)$.

$x_1 = -3$ and $y_1 = 10 \implies$

$$y - 10 = -\frac{3}{2}(x - (-3))$$

$$y - 10 = -\frac{3}{2}(x + 3)$$

...and **Rearrange** this to get the other two forms:

For the form $y = mx + c$, take everything except the y over to the right.

$$y - 10 = -\frac{3}{2}(x + 3)$$

$$\Rightarrow y = -\frac{3}{2}x - \frac{9}{2} + 10$$

$$\Rightarrow y = -\frac{3}{2}x + \frac{11}{2}$$

Equations of Lines

1) **LABEL** the points (x_1, y_1) and (x_2, y_2).

2) **GRADIENT** — find it and call it m.

3) **WRITE DOWN THE EQUATION** using $y - y_1 = m(x - x_1)$

4) **CONVERT** to one of the other forms, if necessary.

To find the form $ax + by + c = 0$, take everything over to one side — and then get rid of any fractions.

Multiply the whole equation by 2 to get rid of the 2's on the bottom line.

$$y = -\frac{3}{2}x + \frac{11}{2}$$

$$\Rightarrow \frac{3}{2}x + y - \frac{11}{2} = 0$$

$$\Rightarrow 3x + 2y - 11 = 0$$

If you end up with an equation like $\frac{3}{2}x - \frac{4}{3}y + 6 = 0$, where you've got a 2 and a 3 on the bottom of the fractions — multiply everything by the <u>lowest common multiple</u> of 2 and 3, i.e. 6.

There ain't nuffink to this geometry lark, Mister...

This is the sort of stuff that looks hard but is actually pretty easy. Finding the equation of a line in that first form really is a piece of cake — the only thing you have to be careful of is when a point has a negative coordinate (or two). In that case, you've just got to make sure you do the subtractions properly when you work out the gradient. See, this stuff ain't so bad...

Coordinate Geometry

More simple stuff for you to have a go at. It's all stuff you've done before, but this time it's used in a different way.

Find the midpoint by **Averaging** each of the coordinates

Don't complain. It doesn't get any easier than this.

Example: Find the midpoint of AB, where A and B are (–3, 10) and (1, 4) respectively.

Find the <u>midpoint</u> by taking the <u>average</u> of the x- and y-coordinates:

Label the points (x_1, y_1) and (x_2, y_2).

Average x-coordinate = $\dfrac{x_1 + x_2}{2} = \dfrac{-3+1}{2} = -1$

Average y-coordinate = $\dfrac{y_1 + y_2}{2} = \dfrac{10+4}{2} = 7$

These are the midpoint coordinates.

So the midpoint has coordinates $(-1, 7)$

Use **Pythagoras** to find the **Length** of a line segment

Example: Find the length of AB, where A and B are (2, 12) and (6, 7) respectively.

Find the <u>length</u> by treating the line segment as the <u>hypotenuse</u> of a right-angled triangle.

Label the points (x_1, y_1) and (x_2, y_2).

Length of side "x" of the triangle = $x_2 - x_1 = 6 - 2 = 4$

Length of side "y" of the triangle = $y_2 - y_1 = 7 - 12 = -5$

So, length of line segment = $\sqrt{(-5)^2 + 4^2} = \sqrt{25 + 16} = 6.4$

Finding where lines meet means solving **Simultaneous Equations**

Okay, you can complain now. This is no fun at all.

Two lines...   Line l_1: $5x + 2y - 9 = 0$ or $y = -\frac{5}{2}x + \frac{9}{2}$ Line l_2: $3x + 4y - 4 = 0$ or $y = -\frac{3}{4}x + 1$

Example: Find where the line l_1 meets the line l_2.

$5x + 2y - 9 = 0$ —①

$3x + 4y - 4 = 0$ —②

Finding where the lines meet means solving these simultaneous equations.

$10x + 4y - 18 = 0$ —③ $= 2 \times$①

$7x - 14 = 0$ —③ – ②

$\Rightarrow x = 2$

Putting this back into equation ② then gives...

$(3 \times 2) + 4y - 4 = 0$

$\Rightarrow 6 + 4y - 4 = 0$

$\Rightarrow 4y = -2$

$\Rightarrow y = -\frac{1}{2}$

Can't remember how to do simultaneous equations? Have a look at pages 20 & 21.

So the lines meet at the point $\left(2, -\frac{1}{2}\right)$.

If you've got the equations in the form y=mx+c — make the right-hand sides of both equations <u>equal</u>.

Line l_1: $y = -\frac{5}{2}x + \frac{9}{2}$ Line l_2: $y = -\frac{3}{4}x + 1$

$-\frac{5}{2}x + \frac{9}{2} = -\frac{3}{4}x + 1$

Solve this equation to find a value for x.

$\Rightarrow -\frac{7}{4}x = -\frac{7}{2}$

$\Rightarrow x = 2$

Then put this value of x into one of the equations to find the y-coordinate...

$y = -\frac{5}{2} \times 2 + \frac{9}{2}$

It doesn't matter which of the equations you use.

$y = -\frac{1}{2}$

So the lines meet at the point $\left(2, -\frac{1}{2}\right)$.

And I think to myself — what a wonderful page...

What an absolutely superb page. There it is, above all these words that you never read. It's fuller than a student at an all-you-can-eat curry house — absolutely jam- (or madras-) packed with useful things about simultaneous equations, lengths and midpoints. Learn this lot, get a few more marks, get the grades you need, and get yourself into some more all-you-can-eat curry houses.

Coordinate Geometry

This page is based around two really important facts that you've got to know — one about parallel lines, one about perpendicular lines. It's really a page of unparalleled excitement...

Two more lines...

Line l_1
$3x - 4y - 7 = 0$
$y = \frac{3}{4}x - \frac{7}{4}$

Line l_2
$x - 3y - 3 = 0$
$y = \frac{1}{3}x - 1$

...and two points...

Point A $(3, -1)$
Point B $(-2, 4)$

Parallel lines have equal Gradient

That's what makes them parallel — the fact that the gradients are the same.

Example: Find the line parallel to l_1 that passes through the point A $(3, -1)$.

Parallel lines have the <u>same gradient</u>.

The original equation is this: $y = \frac{3}{4}x - \frac{7}{4}$

So the new equation will be this: $y = \frac{3}{4}x + c$

We know that the line passes through A, so at this point x will be 3, and y will be –1.

We just need to find c.

Stick these values into the equation to find c.

$$-1 = \frac{3}{4} \times 3 + c$$

$$\Rightarrow c = -1 - \frac{9}{4} = -\frac{13}{4}$$

So the equation of the line is... $y = \frac{3}{4}x - \frac{13}{4}$

And if you're only given the ax + by + c = 0 form it's even easier:

The <u>original</u> line is: $3x - 4y - 7 = 0$

So the <u>new</u> line is: $3x - 4y - k = 0$

Then just use the values of x and y at the point A to find k...

$$3 \times 3 - 4 \times (-1) - k = 0$$

$$\Rightarrow 13 - k = 0$$

$$\Rightarrow k = 13$$

So the equation is: $3x - 4y - 13 = 0$

The gradient of a Perpendicular line is: –1 ÷ the Other Gradient

Finding <u>perpendicular</u> lines (or '<u>normals</u>') is just as easy as finding parallel lines — as long as you remember the gradient of the perpendicular line is <u>–1 ÷ the gradient of the other one</u>.

Example: Find the line perpendicular to l_2 that passes through the point B $(-2, 4)$.

l_2 has equation: $y = \frac{1}{3}x - 1$

So if the equation of the new line is y=mx+c, then

$$m = -1 \div \frac{1}{3}$$

$$\Rightarrow m = -3$$

Since the gradient of a perpendicular line is: –1 ÷ the other one.

Also...

$$4 = (-3) \times (-2) + c$$

$$\Rightarrow c = 4 - 6 = -2$$

Putting the coordinates of B(–2, 4) into y = mx + c.

So the equation of the line is...

$$y = -3x - 2$$

Or if you start with: l_2 $x - 3y - 3 = 0$

To find a perpendicular line, swap these two numbers around, and change the sign of <u>one of them</u>. (So here, 1 and –3 become 3 and 1.)

So the new line has equation...

$$3x + y + d = 0$$

Or you could have used – $3x - y + d = 0$.

But...

$$3 \times (-2) + 4 + d = 0$$

$$\Rightarrow d = 2$$

Using the coordinates of point B.

And so the equation of the <u>perpendicular</u> line is...

$$3x + y + 2 = 0$$

Wowzers — parallel lines on the same graph dimension...

This looks more complicated than it actually is, all this tangent and normal business. All you're doing is finding the equation of a straight line through a certain point — the only added complication is that you have to find the gradient first. And there's another way to remember how to find the gradient of a normal — just remember that the gradients of perpendicular lines multiply together to make –1.

Circles

I always say a beautiful shape deserves a beautiful formula, and here you've got one of my favourite double-acts...

Equation of a circle: $(x - a)^2 + (y - b)^2 = r^2$

The equation of a circle looks complicated, but it's all based on Pythagoras' theorem.
Take a look at the circle below, with centre (6, 4) and radius 3.

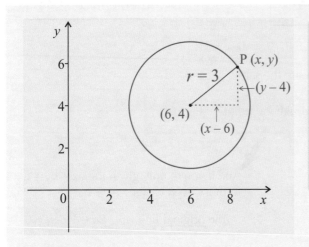

Joining a point P (x, y) on the circumference of the circle to its centre (6, 4), we can create a <u>right-angled triangle</u>.

Now let's see what happens if we use <u>Pythagoras' theorem</u>:

$$(x - 6)^2 + (y - 4)^2 = 3^2$$

or: $(x - 6)^2 + (y - 4)^2 = 9$

This is the equation for the circle. It's as easy as that.

In general, a circle with radius r and centre (a, b) has the equation: $\boxed{(x - a)^2 + (y - b)^2 = r^2}$

This formula is <u>not</u> in the formula book.

Example:

i) What is the centre and radius of the circle with equation $(x - 2)^2 + (y + 3)^2 = 16$

ii) Write down the equation of the circle with centre (–4, 2) and radius 6.

Solution:

i) Comparing $(x - 2)^2 + (y + 3)^2 = 16$ with the general form:

$$(x - a)^2 + (y - b)^2 = r^2$$

then $a = 2$, $b = -3$ and $r = 4$.

> **So the centre (a, b) is: $(2, -3)$**
> **and the radius (r) is: 4.**

And as if by magic, here it is.

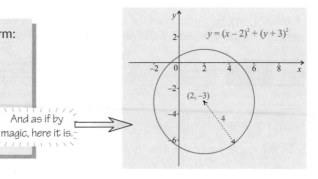

ii) The question says, 'Write down...', so you know you don't need to do any working.
The centre of the circle is (–4, 2), so $a = -4$ and $b = 2$.
The radius is 6, so $r = 6$.
Using the general equation for a circle $(x - a)^2 + (y - b)^2 = r^2$
you can write: $\boxed{(x + 4)^2 + (y - 2)^2 = 36}$

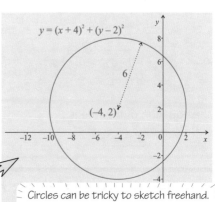

Circles can be tricky to sketch freehand. A compass is definitely the way forward.

This is pretty much all you need to learn. Everything on the next page uses stuff you should know already.

Circles

Rearrange the equation into the *familiar form*

Sometimes you'll be given an equation for a circle that doesn't look much like $(x - a)^2 + (y - b)^2 = r^2$.
This is a bit of a pain, because it means you can't immediately tell what the **radius** is or where the **centre** is.
But all it takes is a bit of **rearranging**.

Let's take the equation: $x^2 + y^2 - 6x + 4y + 4 = 0$

You need to get it into the form $(x - a)^2 + (y - b)^2 = r^2$

This is just like completing the square.

Have a look at pages 12-13 for more on completing the square.

$x^2 + y^2 - 6x + 4y + 4 = 0$

$x^2 - 6x + y^2 + 4y + 4 = 0$

$(x - 3)^2 - 9 + (y + 2)^2 - 4 + 4 = 0$

$(x - 3)^2 + (y + 2)^2 = 9 \implies$ This is the recognisable form, so the centre is **(3, –2)** and the radius is $\sqrt{9} = \mathbf{3}$.

Don't forget the Properties of Circles

You will have seen the circle rules at GCSE. You'll sometimes need to dredge them up in your memory for these circle questions. Here's a reminder of a few useful ones.

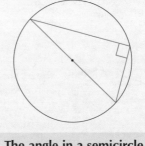

The angle in a semicircle is a right angle.

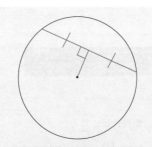

The perpendicular from the centre to a chord bisects the chord.

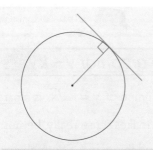

A radius and tangent to the same point will meet at right angles.

Use the Gradient Rule for Perpendicular Lines

Remember that the tangent at a given point will be perpendicular to the radius and the normal at that same point.

Example:
Point A (6, 4) lies on a circle with the equation $x^2 + y^2 - 4x - 2y - 20 = 0$.
  i)   Find the centre and radius of the circle.
  ii) Find the equation of the tangent to the circle at A.

Solution:

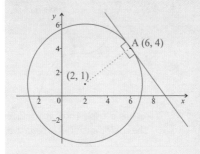

i)   Rearrange the equation to show it as the sum of 2 squares:

$x^2 + y^2 - 4x - 2y - 20 = 0$

$x^2 - 4x + y^2 - 2y - 20 = 0$

$(x - 2)^2 - 4 + (y - 1)^2 - 1 - 20 = 0$

$(x - 2)^2 + (y - 1)^2 = 25$

This shows the centre is (2, 1) and the radius is 5.

ii) The tangent is at right angles to the radius at (6, 4).

Gradient of radius at (6, 4) = $\dfrac{4-1}{6-2} = \dfrac{3}{4}$

Gradient of tangent = $\dfrac{-1}{\frac{3}{4}} = -\dfrac{4}{3}$

Using $y - y_1 = m(x - x_1)$

$y - 4 = -\dfrac{4}{3}(x - 6)$

$3y - 12 = -4x + 24$

$3y + 4x - 36 = 0$

So the chicken comes from the egg, and the egg comes from the chicken...

Well folks, at least it makes a change from all those straight lines and quadratics.
I reckon if you know the **formula** and **what it means**, you should be absolutely **fine** with questions on circles.

Curve Sketching

A picture speaks a thousand words... and graphs are what pass for pictures in maths. They're dead useful in getting your head round tricky questions, and time spent learning how to sketch graphs is time well spent.

The graph of $y = kx^n$ is a different shape for different k and n

Usually, you only need a rough sketch of a graph — so just knowing the basic shapes of these graphs will do.

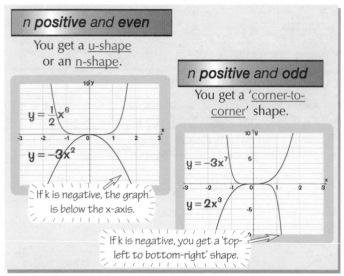

n positive and even

You get a u-shape or an n-shape.

$y = \frac{1}{2}x^6$

$y = -3x^2$

If k is negative, the graph is below the x-axis.

n positive and odd

You get a 'corner-to-corner' shape.

$y = -3x^7$

$y = 2x^3$

If k is negative, you get a 'top-left to bottom-right' shape.

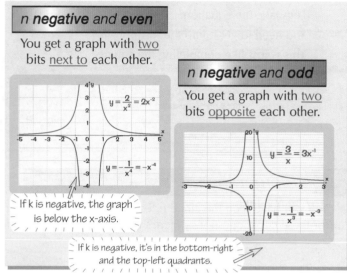

n negative and even

You get a graph with two bits next to each other.

$y = \frac{2}{x^2} = 2x^{-2}$

$y = -\frac{1}{x^4} = -x^{-4}$

If k is negative, the graph is below the x-axis.

n negative and odd

You get a graph with two bits opposite each other.

$y = \frac{3}{x} = 3x^{-1}$

$y = -\frac{1}{x^3} = -x^{-3}$

If k is negative, it's in the bottom-right and the top-left quadrants.

The graph of $y = k\sqrt{x}$ is a Parabola on its Side

The graph of $y = k\sqrt{x}$ is a parabola on its side.

This makes sense really, because if $y = k\sqrt{x}$, then $x = \frac{1}{k^2}y^2$

— and this is just a normal quadratic with the x and y switched round.

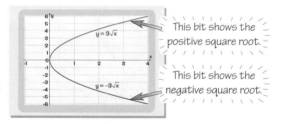

This bit shows the positive square root.

$y = 3\sqrt{x}$

$y = -3\sqrt{x}$

This bit shows the negative square root.

If you know the Factors of a cubic — the graph's easy to Sketch

A cubic function has an x^3 term in it, and all cubics have 'bottom-left to top-right' shape
— or a 'top-left to bottom-right' shape if the coefficient of x^3 is negative.

If you know the factors of a cubic, the graph is easy to sketch — just find where the function is zero.

Example: Sketch the graphs of the following cubic functions.

(i) $f(x) = x(x-1)(2x+1)$ (ii) $g(x) = (1-x)(x^2-2x+2)$ (iii) $h(x) = (x-3)^2(x+1)$ (iv) $m(x) = (2-x)^3$

(i) The function's zero when x = 0, 1 or $-\frac{1}{2}$.

(ii) Differentiate — and you find the gradient's never zero. The coefficient of x^3 is negative, and the quadratic factor of g(x) has no roots — so g(x) is only zero once.

(iii) This has a 'double-root' at x = 3, so the graph just touches the x-axis there but doesn't go through.

(iv) A triple-root looks like this. This has a 'triple-root' at x = 2, and the coefficient of x^3 is negative.

Graphs, graphs, graphs — you can never have too many graphs...

It may seem like a lot to remember, but graphs can really help you get your head round a question — a quick sketch can throw a helluva lot of light on a problem that's got you completely stumped. So being able to draw these graphs won't just help with an actual graph-sketching question — it could help with loads of others too. Got to be worth learning.

Graph Transformations

Suppose you start with any old function f(x). Then you can _transform_ (change) it in three ways — by _translating_ it, _stretching_ or _reflecting_ it.

$$y = f(x)$$

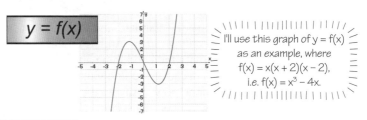

I'll use this graph of y = f(x) as an example, where f(x) = x(x + 2)(x − 2), i.e. f(x) = x³ − 4x.

Translations are caused by Adding things

$y = f(x)+a$

Adding a number to the <u>whole function</u> shifts the graph <u>up or down</u>.

1) If a > 0, the graph goes <u>upwards</u>.
2) If a < 0, the graph goes <u>downwards</u>.

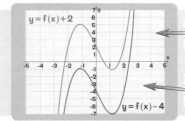

The green graph is y = x(x + 2)(x − 2) + 2, i.e. y = x³ − 4x + 2.

The blue graph is y = x(x + 2)(x − 2) − 4, i.e. y = x³ − 4x − 4.

$y = f(x+a)$

Writing 'x + a' instead of 'x' means the graph moves <u>sideways</u>.

1) If a > 0, the graph goes to the <u>left</u>.
2) If a < 0, the graph goes to the <u>right</u>.

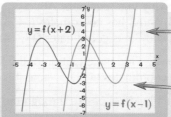

The green graph is y = (x − 1)³ − 4(x − 1), i.e. y = x³ − 3x² − x + 3.

The blue graph is y = (x + 2)³ − 4(x + 2), i.e. y = x³ + 6x² + 8x.

Stretches and Reflections are caused by Multiplying things

$y = af(x)$

<u>Multiplying</u> the <u>whole function</u> <u>stretches</u>, <u>squeezes</u> or <u>reflects</u> the graph <u>vertically</u>.

1) <u>Negative</u> values of 'a' <u>reflect</u> the basic shape in the <u>x-axis</u>.
2) If a > 1 or a < −1 (i.e. |a| > 1) the graph is <u>stretched vertically</u>.
3) If −1 < a < 1 (i.e. |a| < 1) the graph is <u>squashed vertically</u>.

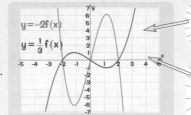

The green graph is y = −2x(x + 2)(x − 2), i.e. y = −2x³ + 8x.

The blue graph is $y = \frac{1}{3}x(x+2)(x-2)$, i.e. $y = \frac{1}{3}x^3 - \frac{4}{3}x$.

$y = f(ax)$

Writing 'ax' instead of 'x' <u>stretches</u>, <u>squeezes</u> or <u>reflects</u> the graph <u>horizontally</u>.

1) <u>Negative</u> values of 'a' <u>reflect</u> the basic shape in the <u>y-axis</u>.
2) If a > 1 or a < −1 (i.e. if |a| > 1) the graph is <u>squashed horizontally</u>.
3) If −1 < a < 1 (i.e. if |a| < 1) the graph is <u>stretched horizontally</u>.

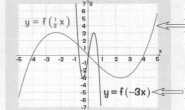

The green graph is $y = \frac{x}{2}\left(\frac{x}{2}+2\right)\left(\frac{x}{2}-2\right)$, i.e. $y = \frac{x^3}{8} - 2x$.

The blue graph is y = −3x(−3x + 2)(−3x − 2), i.e. y = −27x³ + 12x.

More than one transformation at a time: $y = af(bx + c)+d$

Example: $y = af(bx + c) + d$

$a = \frac{1}{3}$ ⇒ The graph is squashed vertically.

$b = \frac{1}{2}$ ⇒ The graph is stretched horizontally.

$c = \frac{1}{2}$ ⇒ The graph is moved horizontally (to the left).

$d = 1$ ⇒ The graph is shifted vertically (upwards).

This is a <u>combination</u> of all these transformations together.

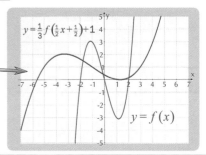

Section Four Revision Questions

There you go then... a section on various geometrical things. And in a way it was quite exciting, I'm sure you'll agree. Because it was in this section that the standard and (some might say) slightly dull straight lines and parabolas were joined by the (ever so slightly) more exciting circle. And as you are probably aware, we mathematicians take our excitement from wherever we can get it. That's the good thing about AS maths really — it teaches you to really look hard for excitement at all opportunities, because you know that it's not going to come around too often. Anyway, that's quite enough of me. I'll leave you alone now to savour the lovely questions below to see how much knowledge you've absorbed as a result of working through the section. If you get them all correct, give yourself a pat on the back. If not, read the section again until you know where you went wrong, and try the questions again.

1) Find the equations of the straight lines that pass through the points

   a) $(2, -1)$ and $(-4, -19)$,   b) $(0, -\frac{1}{3})$ and $(5, \frac{2}{3})$.

   Write each of them in the forms
     i) $y - y_1 = m(x - x_1)$ ii) $y = mx + c$ iii) $ax + by + c = 0$, where a, b and c are integers.

2) Find the point that lies midway between:
     a) $(3, -1)$ and $(-4, 3)$ b) $(10, 4)$ and $(2, 11)$ c) $(96, 9)$ and $(103, 8)$

3) Find the exact lengths of the following line segments, AB:
   a) where A has coordinates $(8, -3)$ and B has coordinates $(-4, 3)$,
   b) where A has coordinates $(3, 4)$ and B has coordinates $(-3, -4)$.

4) a) The line l has equation $y = \frac{3}{2}x - \frac{2}{3}$. Find the equation of the lovely, cuddly line parallel to l, passing through the point with coordinates $(4, 2)$. Name this line Lilly.

   b) The line m (whose name is actually Mike) passes through the point $(6, 1)$ and is perpendicular to $2x - y - 7 = 0$. What is the equation of m?

5) The points A, B and C have coordinates $(1, 4)$, $(4, 5)$ and $(3, 9)$ respectively, and D is the midpoint of BC. Find the equation of the line passing through points A and D.

6) The coordinates of points R and S are $(1, 10)$ and $(9, 3)$ respectively. Find the equation of the line perpendicular to RS, passing through the point midway between them.

7) Give the radius and the coordinates of the centre of the circles with the following equations:
   a) $x^2 + y^2 = 9$
   b) $(x - 2)^2 + (y + 4)^2 = 4$
   c) $x(x + 6) = y(8 - y)$

8) It's lovely, lovely curve sketching time — so draw rough sketches of the following curves:

   a) $y = -2x^4$, b) $y = \frac{7}{x^2}$, c) $y = -5x^3$, d) $y = 2\sqrt{x}$, e) $y = -\frac{2}{x^5}$.

9) Admit it — you love curve-sketching. We all do — and like me, you probably can't get enough of it. So more power to your elbow, and sketch these cubic graphs:

   a) $y = (x - 4)^3$, b) $y = (3 - x)(x + 2)^2$, c) $y = (1 - x)(x^2 - 6x + 8)$, d) $y = (x - 1)(x - 2)(x - 3)$.

10) Right — now it's time to get serious. Put your thinking head on, and use the graph of f(x) to sketch what these graphs would look like after they've been 'transformed'.

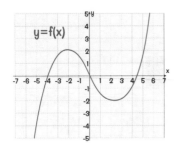

  a) $y = f(ax)$,   where (i) $a > 1$,
                              (ii) $0 < a < 1$,

  b) $y = af(x)$,   where (i) $a > 1$,
                              (ii) $0 < a < 1$,

  c) (i) $y = f(x + a)$, (ii) $y = f(x - a)$,   where $a > 0$,

  d) (i) $y = f(x) + a$, (ii) $y = f(x) - a$,   where $a > 0$.

Differentiation

Brrrrr... differentiation is a bad one — it really is. Not because it's that hard, but because it comes up all over the place in exams. So if you don't know it perfectly, you're asking for trouble.
<u>Differentiation</u> is just a way to work out <u>gradients</u> of graphs. You take a function, differentiate it, and you can quickly tell <u>how steep</u> a graph is. It's magic.

Derivative just means 'the thing you get when you differentiate something'.

$$\frac{d}{dx}\left(x^n\right) = nx^{n-1}$$

$\frac{d}{dx}$ just means 'the derivative of the thing in the brackets'.

Learn this — it won't be in the formula book come the exam.

Use this formula to differentiate **Powers of x**

Equations are much easier to differentiate when they're written as <u>powers of x</u> — like writing $\sqrt{x}$ as $x^{\frac{1}{2}}$.

When you've done this, you can use the formula (the thing in the red box above) to differentiate the equation.

Use the differentiation formula...

For '<u>normal</u>' powers, e.g. x^2

$$y = x^2$$

n is just the power of x.

Here, $n = 2$

So $\dfrac{dy}{dx} = nx^{n-1} = 2x^1 = 2x$

See page 6 for more on rearranging fractions.

For <u>negative</u> powers, e.g. $\dfrac{1}{x^2} = x^{-2}$

$$y = \frac{1}{x^2} = x^{-2}$$

Remember to rewrite the equation as a <u>power</u> of x...

Here, $n = -2$

So $\dfrac{dy}{dx} = nx^{n-1} = -2x^{-3} = -\dfrac{2}{x^3}$

...then use the formula to find the derivative.

For <u>rational</u> powers, e.g. $\sqrt{x} = x^{\frac{1}{2}}$

$$y = \sqrt{x} = x^{\frac{1}{2}}$$

Write the square root as a <u>power</u> of x...

$$n = \frac{1}{2}$$

...and use that very <u>same</u> formula.

$$\frac{dy}{dx} = \frac{1}{2}x^{-\frac{1}{2}} = \frac{1}{2\sqrt{x}}$$

Power Laws:

Differentiation's much easier if you know the Power Laws really well. Like knowing that $x^1 = x$ and $\sqrt{x} = x^{\frac{1}{2}}$.
See page 2 for more info.

Differentiate each term in an equation **Separately**

This formula is better than cake — even better than that really nice sticky black chocolate one from that place in town.
Even if there are loads of terms in the equation, it doesn't matter. Differentiate each bit separately and you'll be fine.

Here are a couple of examples...

If there's a number in front of the function...

$$y = 3\sqrt{x} = 3x^{\frac{1}{2}}$$

$$\frac{dy}{dx} = 3\left(\frac{1}{2}x^{-\frac{1}{2}}\right)$$

...multiply the derivative by the same number.

i.e. $\dfrac{dy}{dx} = \dfrac{3}{2} \times x^{-\frac{1}{2}} = \dfrac{3}{2\sqrt{x}}$

The formula still works with equations like this...

$$y = 6x^2 + \frac{4}{\sqrt[3]{x}} - \frac{2}{x^2} + 1$$

$$= 6x^2 + 4x^{-\frac{1}{3}} - 2x^{-2} + x^0$$

$x^0 = 1$

$$\frac{dy}{dx} = 6(2x) + 4\left(-\frac{1}{3}x^{-\frac{4}{3}}\right) - 2\left(-2x^{-3}\right) + 0x^{-1}$$

Differentiate each bit <u>separately</u>...

$$\frac{dy}{dx} = 12x - \frac{4}{3\sqrt[3]{x^4}} + \frac{4}{x^3}$$

= 0.

...and add or subtract the results.

Dario O'Gradient — differentiating Crewe from the rest...

If you're going to bother doing maths, you've got to be able to differentiate things. Simple as that. But luckily, once you can do the simple stuff, you should be all right. Big long equations are just made up of loads of simple little terms, so they're not really that much harder. Learn the formula, and make sure you can use it by practising all day and all night forever.

Differentiation

Differentiation's what you do if you need to find a gradient. Excited yet?

Differentiate to find Gradients

EXAMPLE: Find the gradient of the graph $y = x^2$ at $x = 1$ and $x = -2$...

You need the gradient of the graph of...

$$y = x^2.$$

So differentiate this function to get...

$$\frac{dy}{dx} = 2x.$$

Now when $x = 1$, $\frac{dy}{dx} = 2$.

And so the gradient of the graph at $x = 1$ is 2.

And when $x = -2$, $\frac{dy}{dx} = -4$.

So the gradient of the graph at $x = -2$ is -4.

Use differentiation to find the gradient of a <u>curve</u> — which is the same as the gradient of the <u>tangent</u> at that point.

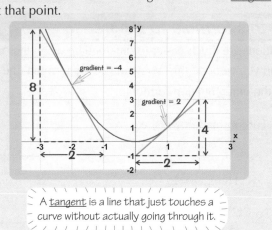

A <u>tangent</u> is a line that just touches a curve without actually going through it.

Find out if a function is Increasing or Decreasing

You can also use differentiation to work out exactly where a function is <u>increasing</u> or <u>decreasing</u> — and how quickly.

A function is <u>increasing</u> when...
...the gradient is <u>positive</u>.

y gets bigger...

...as x gets bigger.

A function is <u>decreasing</u> when...
...the gradient is <u>negative</u>.

y gets smaller...

...as x gets bigger.

The <u>bigger</u> the gradient...
...the <u>faster</u> y changes with x.

A small change in x means a big change in y.

A big change in x means a small change in y.

Example: Find where the following function is <u>increasing</u> and <u>decreasing</u>: $f(x) = 3x^2 - 6x$.

This is a question about <u>gradients</u> — so <u>differentiate</u>.

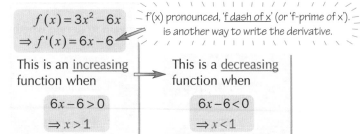

$$f(x) = 3x^2 - 6x$$
$$\Rightarrow f'(x) = 6x - 6$$

f'(x) pronounced, 'f dash of x' (or 'f-prime of x'). is another way to write the derivative.

This is an <u>increasing</u> function when

$$6x - 6 > 0$$
$$\Rightarrow x > 1$$

This is a <u>decreasing</u> function when

$$6x - 6 < 0$$
$$\Rightarrow x < 1$$

Differentiation and Gradients

To find the gradient of a curve at a certain point:

1) <u>Differentiate the equation</u> of the curve.

2) <u>Work out the derivative</u> at the point.

An increasing function has a <u>positive</u> gradient.

A decreasing function has a <u>negative</u> gradient.

Help me Differentiation — You're my only hope...

There's not much hard maths on this page — but there are a couple of very important ideas that you need to get your head round pretty darn soon. Understanding that differentiating gives the gradient of the graph is more important than washing regularly — AND THAT'S IMPORTANT. The other thing on the page — that you can tell whether a function is getting bigger or smaller by looking at the derivative — is also vital. Sometimes the examiners ask you to find where a function is increasing or decreasing — so you'd just have to find where the derivative was positive or negative.

Differentiation

To find a <u>stationary point</u>, you need to find where the graph 'levels off' — that means where the <u>gradient</u> becomes <u>zero</u>.

Stationary Points are when the gradient is Zero

EXAMPLE: Find the stationary points on the curve $y = 2x^3 - 3x^2 - 12x + 5$, and work out the nature of each one.

A <u>stationary point</u> can be...

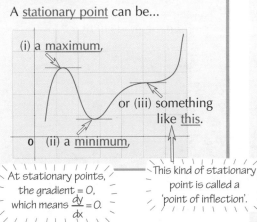

(i) a <u>maximum</u>,

or (iii) something like <u>this</u>.

0   (ii) a <u>minimum</u>,

At stationary points, the gradient = 0, which means $\frac{dy}{dx} = 0$.

This kind of stationary point is called a 'point of inflection'.

A turning point is a another name for a maximum or a minimum.

You need to find where $\frac{dy}{dx} = 0$. So first, <u>differentiate</u> the function.

$$y = 2x^3 - 3x^2 - 12x + 5$$
$$\Rightarrow \frac{dy}{dx} = 6x^2 - 6x - 12$$

This is the expression for the gradient.

And then set this derivative equal to <u>zero</u>.

$$6x^2 - 6x - 12 = 0$$
$$\Rightarrow x^2 - x - 2 = 0$$
$$\Rightarrow (x-2)(x+1) = 0$$
$$\Rightarrow x = 2 \text{ or } x = -1$$

See pages 10 to 14 for more about solving quadratics.

So the graph has <u>two</u> stationary points, at $x = 2$ and $x = -1$.

Decide if it's a Maximum or a Minimum by differentiating Again

Once you've found where the stationary points are, you have to decide whether each of them is a <u>maximum</u> or <u>minimum</u> — this is all a question means when it says, '...determine the <u>nature</u> of the turning points'.

To decide whether a stationary point is a <u>maximum</u> or a <u>minimum</u> — just differentiate again to find $\frac{d^2y}{dx^2}$.

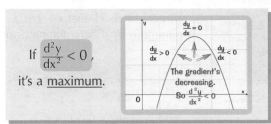

If $\frac{d^2y}{dx^2} < 0$, it's a <u>maximum</u>.

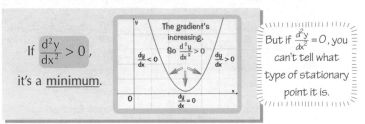

If $\frac{d^2y}{dx^2} > 0$, it's a <u>minimum</u>.

But if $\frac{d^2y}{dx^2} = 0$, you can't tell what type of stationary point it is.

You've just found that $\frac{dy}{dx} = 6x^2 - 6x - 12$.

Stick in the x-coordinates of the stationary points.

So differentiating again gives $\frac{d^2y}{dx^2} = 12x - 6$.

At $x = -1$, $\frac{d^2y}{dx^2} = -18$, which is <u>negative</u> — so $x = -1$ is a <u>maximum</u>.

And at $x = 2$, $\frac{d^2y}{dx^2} = 18$, which is <u>positive</u> — so $x = 2$ is a <u>minimum</u>.

And since a cubic graph (where the coefficient of x^3 is <u>positive</u>) goes from <u>bottom-left to top-right</u>...

...you can draw a rough sketch of the graph, even though the roots would be hard to find.

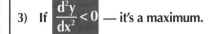

Stationary Points

1) **Find stationary points by solving** $\frac{dy}{dx} = 0$.

2) **Differentiate again to decide whether this is a maximum or a minimum.**

3) **If** $\frac{d^2y}{dx^2} < 0$ — **it's a maximum.**

   If $\frac{d^2y}{dx^2} > 0$ — **it's a minimum.**

An anagram of differentiation is "Perfect Insomnia Cure"...

No joke this, is it — this differentiation business — but it's a dead important topic in maths. It's so important to know how to find whether a stationary point is a max or a min — but it can get a bit confusing. Try remembering MINMAX — which is short for 'MINUS means a MAXIMUM'. Or make up some other clever way to remember what means what.

Curve Sketching

You'll even be asked to do some drawing in the exam... but don't get too excited — it's just drawing graphs... great.

Find where the curve crosses the Axes

Sketch the graph of $f(x) = \frac{x^2}{2} - 2\sqrt{x}$, for $x \geq 0$.

The curve crosses the y-axis when $x = 0$ — so put $x = 0$ in the expression for y.

When $x = 0$, $f(x) = 0$ — and so the curve goes through the origin.

The curve crosses the x-axis when $f(x) = 0$. So solve

$$\frac{x^2}{2} - 2\sqrt{x} = 0$$

$$\Rightarrow x^2 = 4\sqrt{x} = 4x^{\frac{1}{2}}$$

> Dividing both sides by $x^{\frac{1}{2}}$.

$$\Rightarrow x^{\frac{3}{2}} = 4$$

$$\Rightarrow x = 4^{\frac{2}{3}} = \sqrt[3]{4^2} = \sqrt[3]{16} \approx 2.5$$

And so the curve crosses the x-axis when $x \approx 2.5$.

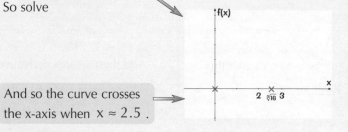

...Differentiate to find Gradient info...

Differentiating the function gives...

$$f(x) = \frac{1}{2}x^2 - 2x^{\frac{1}{2}}$$

$$\Rightarrow f'(x) = \frac{1}{2}(2x) - 2\left(\frac{1}{2}x^{-\frac{1}{2}}\right) = x - x^{-\frac{1}{2}} = x - \frac{1}{\sqrt{x}}$$

> Using the derivative — you can find stationary points and tell when the graph goes 'uphill' and 'downhill'.

> This is the quickest way to check if something's a max or a min.

1) So there's a stationary point when...

$$x - \frac{1}{\sqrt{x}} = 0$$

$$\Rightarrow x = \frac{1}{\sqrt{x}}$$

$$\Rightarrow x^{\frac{3}{2}} = 1 \Rightarrow x = 1$$

And at $x = 1$, $f(x) = \frac{1}{2} - 2 = -\frac{3}{2}$.

2) The gradient's negative when...

$$x - \frac{1}{\sqrt{x}} < 0$$

$$\Rightarrow x < \frac{1}{\sqrt{x}}$$

$$\Rightarrow x^{\frac{3}{2}} < 1 \Rightarrow x < 1$$

> So the function decreases when $0 \leq x < 1$...

3) The gradient's positive when...

$$x - \frac{1}{\sqrt{x}} > 0$$

$$\Rightarrow x > 1$$

> ...and increases for $x > 1$.

You could check that $x = 1$ is a minimum by differentiating again.

$$f''(x) = 1 - \left(-\frac{1}{2}x^{-\frac{3}{2}}\right) = 1 + \frac{1}{2\sqrt{x^3}}$$

This is positive when $x = 1$, and so this is definitely a minimum.

...and find out what happens when x gets Big

You can also try and decide what happens as x gets very big — in both the positive and negative directions.

Factorise f(x) by taking the biggest power outside the brackets...

$$\frac{x^2}{2} - 2\sqrt{x} = x^2\left(\frac{1}{2} - 2x^{-\frac{3}{2}}\right) = x^2\left(\frac{1}{2} - \frac{2}{x^{\frac{3}{2}}}\right)$$

> As x gets large, this bit disappears — and the bit in brackets gets closer to $\frac{1}{2}$.

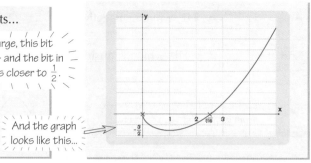

> And the graph looks like this...

And so as x gets larger, f(x) gets closer and closer to $\frac{1}{2}x^2$ — and this just keeps growing and growing.

Curve-sketching's important — but don't take my word for it...

Curve-sketching — an underrated skill, in my opinion. As Shakespeare once wrote, 'Those who can do fab sketches of graphs and stuff are likely to get pretty good grades in maths exams, no word of a lie'. Well, he probably would've written something like that if he was into maths. And he would've written it because graphs are helpful when you're trying to work out what a question's all about — and once you know that, you can decide the best way forward. And if you don't believe me, remember the saying of the ancient Roman Emperor Caesar, 'If in doubt, draw a graph'.

Finding Tangents and Normals

What's a tangent? Beats me. Oh no, I remember, it's one of those thingies on a curve. Ah, yes... I remember now...

Tangents *Just* touch a curve

To find the equation of a tangent or a normal to a curve, you first need to know its gradient — so differentiate. Then complete the line's equation using the coordinates of one point on the line.

Find the tangent to the curve $y = (4 - x)(x + 2)$ at the point (2, 8).

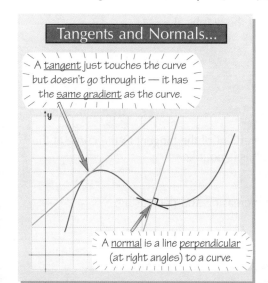

Tangents and Normals...

A tangent just touches the curve but doesn't go through it — it has the same gradient as the curve.

A normal is a line perpendicular (at right angles) to a curve.

To find the curve's (and the tangent's) gradient, first write the equation in a form you can differentiate...

$$y = 8 + 2x - x^2$$

...and then differentiate it.

$$\frac{dy}{dx} = 2 - 2x$$

The gradient of the tangent will be the gradient of the curve at $x = 2$.

At $x = 2$, $\frac{dy}{dx} = -2$,

So the tangent has equation,

$$y - y_1 = -2(x - x_1)$$

in $y - y_1 = m(x - x_1)$ form. See page 25.

And since it passes through the point (2,8), this becomes

$$y - 8 = -2(x - 2), \text{ or } y = -2x + 12.$$

You can also write it in $y = mx + c$ form.

Normals are at **Right Angles** to a curve

EXAMPLE: Find the normal to the curve $y = \dfrac{(x + 2)(x + 4)}{6\sqrt{x}}$ at the point (4, 4).

Write the equation of the curve in a form you can differentiate.

$$y = \frac{x^2 + 6x + 8}{6x^{\frac{1}{2}}} = \frac{1}{6}x^{\frac{3}{2}} + x^{\frac{1}{2}} + \frac{4}{3}x^{-\frac{1}{2}}$$

Dividing everything on the top line by everything on the bottom line.

Differentiate it...

$$\frac{dy}{dx} = \frac{1}{6}\left(\frac{3}{2}x^{\frac{1}{2}}\right) + \frac{1}{2}x^{-\frac{1}{2}} + \frac{4}{3}\left(-\frac{1}{2}x^{-\frac{3}{2}}\right)$$

$$= \frac{1}{4}\sqrt{x} + \frac{1}{2\sqrt{x}} - \frac{2}{3\sqrt{x^3}}$$

Find the gradient at the point you're interested in. At $x = 4$,

$$\frac{dy}{dx} = \frac{1}{4} \times 2 + \frac{1}{2 \times 2} - \frac{2}{3 \times 8} = \frac{2}{3}$$

Because the gradient of the normal multiplied by the gradient of the curve must be −1.

So the gradient of the normal is $-\frac{3}{2}$.

And the equation of the normal is $y - y_1 = -\frac{3}{2}(x - x_1)$.

Finally, since the normal goes through the point (4, 4), the equation of the normal must be $y - 4 = -\frac{3}{2}(x - 4)$, or after rearranging, $y = -\frac{3}{2}x + 10$.

Finding tangents and normals

1) **Differentiate the function.**

2) **Find the gradient, m, of the tangent or normal. This is,**

   for a tangent: the gradient of the curve

   for a normal: $\dfrac{-1}{\text{gradient of the curve}}$

3) **Write the equation of the tangent or normal in the form $y - y_1 = m(x - x_1)$, or $y = mx + c$.**

4) **Complete the equation of the line using the coordinates of a point on the line.**

Repeat after me... "I adore tangents and normals..."

Examiners can't stop themselves saying the words 'Find the tangent...' and 'Find the normal...'. They love the words. These phrases are music to their ears. They can't get enough of them. I just thought it was my duty to tell you that. And so now you know, you'll definitely be wanting to learn how to do the stuff on this page. Of course you will.

Section Five Revision Questions

That's what differentiation is all about. And frankly, it probably isn't the worst topic you'll meet in AS maths. Yes, there are fiddly things to remember — but overall, it's not as bad as all that. And just think of all the lovely marks you'll get if you can answer questions like these in the exam...

1) An easy one to start with. Write down the formula for differentiating any power of x.

2) Differentiate these functions with respect to x, and then find the gradients of the graphs at x = 1:
   a) $y = x^2 + 2$,
   b) $y = x^4 + \sqrt{x}$,
   c) $y = \frac{7}{x^2} - \frac{3}{\sqrt{x}} + 12x^3$

3) What's the connection between the gradient of a curve at a point and the gradient of the tangent to the curve at the same point? (That sounds like a joke in need of a punchline — but sadly, this is no joke.)

4) Not so easy — it involves inequalities: find when these two functions are increasing and decreasing:
   a) $y = 6(x+2)(x-3)$,
   b) $y = \frac{1}{x^2}$.

5) Write down what a stationary point is. Find the stationary points of the graph $y = x(x-8)(x-1)$ (to 3 s.f.)

6) How can you decide whether a stationary point is a maximum or a minimum?

7) A bit fiddly this — you won't like it — so just make sure you can do it.
   Find the stationary points of the function $y = x^3 + \frac{3}{x}$.
   Decide whether each stationary point is a minimum or a maximum.

8) Now find the stationary points of the function $y = x^3 - 3x$.
   Again, decide whether each stationary point is a minimum or a maximum.

9) Yawn, yawn. Find the equations of the tangent and the normal to the curve $y = \sqrt{x^3} - 3x - 10$ at x = 16.

10) Show that the lines $y = \frac{x^3}{3} - 2x^2 - 4x + \frac{86}{3}$ and $y = \sqrt{x}$ both go through the point (4,2), and are perpendicular at that point. Good question, that — nice and exciting, just the way you like 'em.

General Certificate of Education
Advanced Subsidiary (AS) and Advanced Level

Core 1 Mathematics — Practice Exam One

You are NOT allowed to use a calculator.

1 (i) Write down the exact positive value of $36^{-\frac{1}{2}}$ [2]

(ii) Simplify $\dfrac{a^6 \times a^3}{\sqrt{a^4}} \div a^{\frac{1}{2}}$ [2]

(iii) Express $\left(5\sqrt{5} + 2\sqrt{3}\right)^2$ in the form $a + b\sqrt{c}$, where a, b and c are integers to be found. [4]

2 (i) Either algebraically, or by sketching the graphs, solve the inequality $4x + 7 > 7x + 4$ [2]
(ii) Find the values of k, such that $(x-5)(x-3) > k$ for all possible values of x. [3]
(iii) Find the range of x that satisfies the inequality $(x+3)(x-2) < 2$. [3]

3 (i) (a) Solve the simultaneous equations
$$y = (x-2)(x-4) \qquad y = 4 - 2x$$ [3]
(b) Interpret your answer to part (a) geometrically. [2]
A function in x is given by f$(x) = x^2 + mx + 25$, where m is a constant.
(ii) (a) Find values of m such that f(x) has no real roots. [2]
(b) Find values of m such that f(x) has just one root, and for each of these values of m,
solve the equation f$(x) = 0$. [3]

4 (i) Find the coordinates of the point A, when A lies at the intersection of the lines l_1 and l_2, and when the
equations of l_1 and l_2 respectively are:
$$x - y + 1 = 0 \text{ and } 2x + y - 8 = 0.$$ [3]

The points B and C have coordinates $(6, -4)$ and $\left(-\dfrac{4}{3}, -\dfrac{1}{3}\right)$ respectively, and D is the midpoint of AC.
(ii) Find the equation of the line BD in the form $ax + by + c = 0$, where a, b and c are integers. [5]
(iii) Show that the triangle ABD is a right-angled triangle. [3]

5 Find dy/dx for each of the following:
(i) $y = x^2$ [1]
(ii) $y = 3x^4 - 2x$ [2]
(iii) $y = (x^2 + 4)(x - 2)$ [2]

6 The diagram shows a circle. A $(2, 1)$ and B $(0, -5)$ lie on the circle and AB is a diameter.
C $(4, -1)$ is also on the circle.
(i) Find the centre and radius of the circle. [4]
(ii) Show that the equation of the circle can be written in the form:
$$x^2 + y^2 - 2x + 4y - 5 = 0$$ [3]
(iii) The tangent at A and the normal at C cross at D.
Find the co-ordinates of D. [8]

7 (i) Express $x^2 - 6x + 5$ in the form $(x + a)^2 + b$. [2]
(ii) Factorise the expression $x^2 - 6x + 5$. [2]
(iii) Hence sketch the graph of $y = x^2 - 6x + 5$, clearly indicating the coordinates
of the vertex and the points where it cuts the axes. [2]

8 The function $g(x)$ is defined by $g(x) = \dfrac{x(3x^2 - 2x - 9)}{\sqrt{x}}$, for $x > 0$.

(i) Show that the derivative of $g(x)$ can be written $g'(x) = \dfrac{3}{2\sqrt{x}}(5x^2 - 2x - 3)$. [3]
(ii) Hence find the x-coordinate of the stationary point ($x > 0$) of $g(x)$. [3]
(iii) By evaluating $g''(x)$ or otherwise, find the nature of this stationary point. [3]

Paper 1 Q1 — Powers and Surds

---

1 (i) Write down the exact positive value of $36^{-\frac{1}{2}}$ [2]

(ii) Simplify $\dfrac{a^6 \times a^3}{\sqrt{a^4}} \div a^{\frac{1}{2}}$ [2]

(iii) Express $\left(5\sqrt{5} + 2\sqrt{3}\right)^2$ in the form $a + b\sqrt{c}$, where a, b and c are integers to be found. [4]

(i) Power Law questions can usually be answered by Rearranging

'Write down the exact value of $36^{-\frac{1}{2}}$.'

With <u>Power Law</u> questions, you usually just have to remember a couple of basic <u>formulas</u>, then do a couple of sums pretty darn <u>carefully</u>. If you've forgotten any of the Laws, most of them are in the boring-looking box at the bottom of the next page. So, on with the question...

First get rid of the minus in the exponent / power...

$$x^{-n} = \frac{1}{x^n} \implies 36^{-\frac{1}{2}} = \frac{1}{36^{\frac{1}{2}}}$$

Then deal with the $\frac{1}{2}$.

$$\frac{1}{36^{\frac{1}{2}}} = \frac{1}{\sqrt{36}} \quad x^{\frac{1}{n}} = \sqrt[n]{x}$$

And finally...

Do a simple square root. $\quad \dfrac{1}{\sqrt{36}} = \dfrac{1}{6}$

The question asks for the positive value, so you don't need to write "or –1/6".

(ii) Simplifying just means Rearranging as well

'Simplify $\dfrac{a^6 \times a^3}{\sqrt{a^4}} \div a^{\frac{1}{2}}$.'

When you're simplifying powers, it's a good idea to get them all looking <u>the same</u>. In this example, get the individual bits in the form a^n.

See page 2 for more info on the Power Laws.

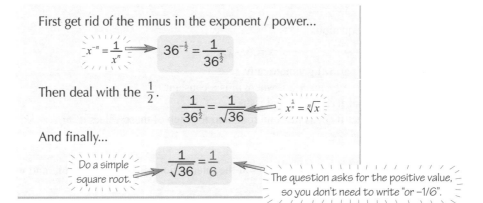

First simplify the tricky bit on the bottom of the fraction...

$$\sqrt[m]{a^n} = a^{\frac{n}{m}} \quad \frac{a^6 \times a^3}{\sqrt{a^4}} \div a^{\frac{1}{2}} = \frac{a^6 \times a^3}{a^2} \div a^{\frac{1}{2}}$$

Then rewrite this so that you're only multiplying things.

$$a^6 \times a^3 \times a^{-2} \times a^{-\frac{1}{2}} \quad \frac{1}{a^n} = a^{-n}$$

Dividing by a^n is the same as multiplying by a^{-n}.

And then just add all the powers together, to get

$$a^6 \times a^3 \times a^{-2} \times a^{-\frac{1}{2}} = a^{6+3-2-\frac{1}{2}}$$

$$= a^{\frac{13}{2}} \quad \text{Hurray...}$$

There are lots of different ways of tackling a question like this. However you do it, don't rush, and make sure you check it through at the end.

This may sound stupid, but questions on Power Laws aren't too bad, as long as you <u>obey the Power Laws</u>. It really is that simple.

Paper 1 Q1 — Powers and Surds

(iii) Multiply out the brackets — then use the rules for Surds

'Express $\left(5\sqrt{5}+2\sqrt{3}\right)^2$ in the form $a+b\sqrt{c}$, where a, b and c are integers to be found.'

Yet again, you've got to simplify and rearrange the equation.

First of all (after the initial shock of "Arrrgghh — too many surds.") you should notice the squared sign around the brackets. And that's the first thing to go...

Multiply out the brackets first.

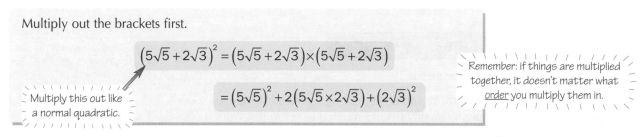

$$\left(5\sqrt{5}+2\sqrt{3}\right)^2=\left(5\sqrt{5}+2\sqrt{3}\right)\times\left(5\sqrt{5}+2\sqrt{3}\right)$$

$$=\left(5\sqrt{5}\right)^2+2\left(5\sqrt{5}\times2\sqrt{3}\right)+\left(2\sqrt{3}\right)^2$$

Multiply this out like a normal quadratic.

Remember: if things are multiplied together, it doesn't matter what order you multiply them in.

This next bit's a tad confusing, I reckon. You've got three terms to deal with, and they're all a little bit nasty.

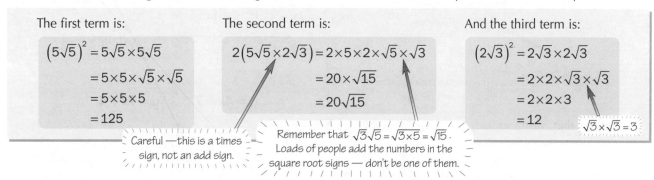

The first term is:

$$\left(5\sqrt{5}\right)^2=5\sqrt{5}\times5\sqrt{5}$$
$$=5\times5\times\sqrt{5}\times\sqrt{5}$$
$$=5\times5\times5$$
$$=125$$

The second term is:

$$2\left(5\sqrt{5}\times2\sqrt{3}\right)=2\times5\times2\times\sqrt{5}\times\sqrt{3}$$
$$=20\times\sqrt{15}$$
$$=20\sqrt{15}$$

And the third term is:

$$\left(2\sqrt{3}\right)^2=2\sqrt{3}\times2\sqrt{3}$$
$$=2\times2\times\sqrt{3}\times\sqrt{3}$$
$$=2\times2\times3$$
$$=12$$

$\sqrt{3}\times\sqrt{3}=3$

Careful —this is a times sign, not an add sign.

Remember that $\sqrt{3}\sqrt{5}=\sqrt{3\times5}=\sqrt{15}$. Loads of people add the numbers in the square root signs — don't be one of them.

Now you've done that, though, you're almost home and dry. All that's left to do is add the three terms together.

So the whole thing's equal to...

$$125+20\sqrt{15}+12$$
$$=137+20\sqrt{15}$$

And since 137, 20 and 15 are all integers, that's the final answer to the question.

The Fabulous Power Laws and the Rules for Surds

Okay, you've seen them often enough — and here they are again.
Maybe they're here again because they're important. Mmm. Could be.

The Forgotten and Magical Power Laws

$$y^m\times y^n=y^{m+n} \qquad y^m\div y^n=y^{m-n} \qquad y^{\frac{n}{m}}=\sqrt[m]{y^n}$$

$$y^{-n}=\frac{1}{y^n} \qquad y^0=1 \qquad \left(y^m\right)^n=y^{mn} \qquad y^1=y$$

Rules of Surds

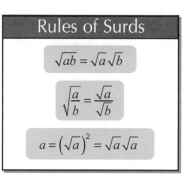

$$\sqrt{ab}=\sqrt{a}\sqrt{b}$$

$$\sqrt{\frac{a}{b}}=\frac{\sqrt{a}}{\sqrt{b}}$$

$$a=\left(\sqrt{a}\right)^2=\sqrt{a}\sqrt{a}$$

Keep your Magic Power Laws close to you at all times...

I don't want to go on and on and start ranting but... hang on a bit, I do want to go on and on, and I do want to rant. The thing is that questions on the Power Laws always come up in the exams and are always worth a good few marks. And those few marks could make the difference between one grade and the next. Think about it — half an hour spent learning this page really well could move you up a grade. You know it makes sense...

Paper 1 Q2 — Inequalities...

> **2 (i)** Either algebraically, or by sketching the graphs, solve the inequality
> $$4x + 7 > 7x + 4$$
> [2]
> **(ii)** Find the values of k, such that
> $$(x-5)(x-3) > k \text{ for all possible values of } x.$$
> [3]
> **(iii)** Find the range of x that satisfies the inequality
> $$(x+3)(x-2) < 2.$$
> [3]

(i) A straightforward Linear Inequality

'Either algebraically, or by sketching the graphs...'

That opening makes it sound really tricky, but don't be fooled. Sketching graphs sounds much easier, but it's such a <u>simple</u> inequality that it's much quicker to just work it out:

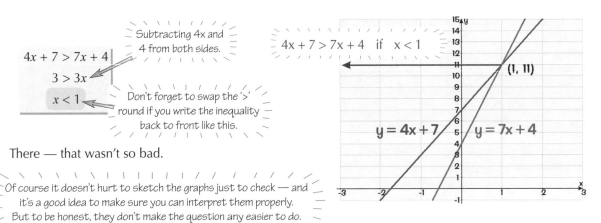

Subtracting 4x and 4 from both sides.

$4x + 7 > 7x + 4$

$3 > 3x$

$x < 1$

Don't forget to swap the '>' round if you write the inequality back to front like this.

$4x + 7 > 7x + 4$ if $x < 1$

$y = 4x + 7$ $y = 7x + 4$ (1, 11)

There — that wasn't so bad.

Of course it doesn't hurt to sketch the graphs just to check — and it's a good idea to make sure you can interpret them properly. But to be honest, they don't make the question any easier to do.

(ii) Easy — if you spot the Symmetry

'Find the values of k, such that
$$(x-5)(x-3) > k \text{ for all possible values of } x.'$$

Basically, you've got to find the <u>minimum</u> value of $(x-5)(x-3)$, and make sure k is less than that.

Another way you could do this part would be to multiply out the brackets and complete the square. You could also differentiate to find the minimum point.

As always, if you're a bit unsure where to start, think what the function looks like — and <u>SKETCH THE GRAPH</u>.

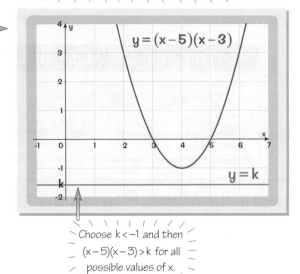

$y = (x-5)(x-3)$

$y = k$

Here's the cunning bit: The thing to realise is that the graph's <u>symmetrical</u> — so the minimum will be halfway between $x = 3$ and $x = 5$ — i.e. $x = 4$. So just plug that into the equation to find the lowest point the graph reaches:

Putting $x = 4$ in $(x-5)(x-3)$ gives
$$(4-5)(4-3) = -1 \times 1$$
$$= -1$$

So if $k < -1$, the graph will never be as low as k.

Choose k < −1 and then $(x-5)(x-3) > k$ for all possible values of x.

Paper 1 Q2 — ...and Quadratics

(iii) Draw the **Graph** to see when it's **Negative**

'Find the range of x that satisfies the inequality

$$(x+3)(x-2)<2 \text{ .'}$$

You need to find a 'range of x' — not just one value. Sounds a bit complicated — but it's not too bad once you get going. The first thing to do is rearrange the equation so you've got <u>zero</u> on one side.

Rearrange the expression to get...

$$(x+3)(x-2)<2$$
$$x^2+x-6<2$$
$$x^2+x-8<0$$

Rearrange this to get zero on one side. Then when you draw the graph, all you need to do is find where the graph is negative.

Then sketch the graph of $y = x^2 + x - 8$. And since you're interested in when this is less than zero, make sure you find out where this crosses the <u>x-axis</u>.

When the graph crosses the x-axis, it changes from positive to negative, or vice versa.

Now $x^2 + x - 8$ doesn't factorise — so find out where it crosses the x-axis by using the <u>quadratic formula</u>.

You can tell it doesn't factorise because $\sqrt{b^2-4ac} = \sqrt{33} = 5.74456...$ — and that's not a whole number or an 'easy' decimal.

Now, $x^2 + x - 8 = 0$ when

$$x = \frac{-1\pm\sqrt{1^2-(4\times1\times-8)}}{2\times1}$$
$$= \frac{-1\pm\sqrt{33}}{2}$$

The quadratic formula:
$$x = \frac{-b\pm\sqrt{b^2-4ac}}{2a}$$
when
$$ax^2+bx+c=0$$

So the graph looks like this:

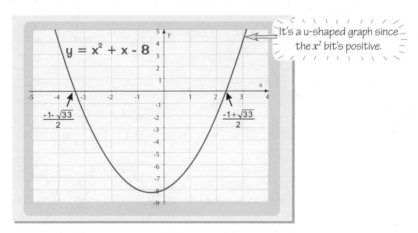

It's a u-shaped graph since the x^2 bit's positive.

And since you need this to be <u>negative</u> — it's pretty clear that the range of x you're interested in is...

$$\frac{-1-\sqrt{33}}{2} < x < \frac{-1+\sqrt{33}}{2}$$

$a < x < b$ means 'x between a and b'.

or... $-3.37 < x < 2.37$ (to 2 d.p.)

It's easy to check — just stick the two numbers back into the original inequality and make sure the left-hand side equals 2.

These can be pretty darn hard unless you draw the graphs...

It's true. These questions can be very hard unless you have a picture to look at. It just helps you understand exactly what's going on, and what the examiners are going on about. That's the thing with these questions — they can look so intimidating. But drawing a picture helps you get your head round it — and once you've got your head round it, it's much easier to work towards the answer. Mmmm... I think I've said enough on that for now.

Paper 1 Q3 — Simultaneous Eqns...

> **3 (i) (a)** Solve the simultaneous equations
>
> $$y = (x-2)(x-4) \qquad\qquad y = 4 - 2x$$ [3]
>
> **(b)** Interpret your answer to part (a) geometrically. [2]
>
> A function in x is given by $f(x) = x^2 + mx + 25$, where m is a constant.
>
> **(ii) (a)** Find values of m such that $f(x)$ has no real roots. [2]
>
> **(b)** Find values of m such that $f(x)$ has just one root, and for each of these values of m,
> solve the equation $f(x) = 0$. [3]

(i)(a) | A nice easy Start — Simultaneous Equations

It's always nice to label them to start with:

$$y = (x-2)(x-4) \quad \text{——①}$$
$$y = 4 - 2x \quad \text{——②}$$

Simultaneous equations where one of them is quadratic — straight away, think <u>substitution</u>. So...

Substitute y from equation 2 into equation 1:

$$4 - 2x = (x-2)(x-4)$$
$$= x^2 - 6x + 8$$

...and rearranging so that everything is on one side gives you...

$$x^2 - 4x + 4 = 0$$
$$\Rightarrow (x-2)^2 = 0$$
$$\Rightarrow (x-2) = 0$$
$$\Rightarrow x = 2$$

Now find y:

$$x = 2 \text{ in equation 2} \Rightarrow y = 4 - 2 \times 2 = 0$$

So there's only one solution — x = 2, y = 0.

(i)(b) | Sketch it

①
$$y = (x-2)(x-4) \quad \longleftarrow \text{ It crosses the x-axis when y = 0, i.e.}$$
$$= x^2 - 6x + 8 \qquad\qquad \text{when either of these brackets is 0.}$$

The graph of the first equation is a u-shaped parabola,
which crosses the x-axis at x = 2 and x = 4.

And if you differentiate this, you find that $\frac{dy}{dx} = 2x - 6$,
and so the gradient at x = 2 is –2.

②
$$y = -2x + 4 \quad \longleftarrow \text{ Here's eqn 2 in y=mx+c form.}$$

The graph of equation 2 is a straight line with gradient –2.
To draw its graph, just find where it crosses the x- and y- axes:
It crosses the x-axis at x=2 and the y-axis at y=4.

③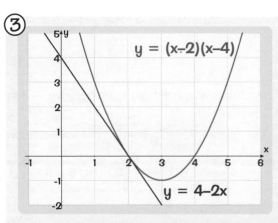

So you can see from the graphs that the straight
line is a <u>tangent</u> to the curve at (2,0).

You should have known this anyway from
the fact that we only got one solution
for part i(a). See page 23 for more info.

Paper 1 Q3 — ...and Quadratics

(ii)(a) | Use the Discriminant

'Find values of m such that $f(x)$ has no real roots.'

1) When a question starts mentioning the **number of roots** of a quadratic, you're bound to need the **discriminant**.

2) There's more about this on page 15, but in short it's 'the bit in the square root' from the quadratic formula:

$$b^2 - 4ac$$

3) The square roots of a negative number aren't real — so to get no real roots you want $b^2 - 4ac < 0$.

For this question, you've got: $a = 1$, $b = m$ and $c = 25$

So for $b^2 - 4ac < 0$, you get:

$$m^2 - 4 \times 1 \times 25 < 0$$
$$m^2 - 100 < 0$$
$$m^2 < 100$$
$$-10 < m < 10$$

Always remember that it's the discriminant that determines how many roots a quadratic has.
If $b^2 - 4ac > 0$, it'll have 2 roots.
If $b^2 - 4ac = 0$, it'll have 1 root.
If $b^2 - 4ac < 0$, it'll have 0 real roots.

(ii)(b) | Guess what — Use the Discriminant again...

'Find values of m such that $f(x)$ has just one root...'

For just one root, you need the discriminant = 0:

$$b^2 - 4ac = 0$$
$$m^2 - 100 = 0$$
$$m^2 = 100$$
$$m = \pm 10$$

'for each of these values of m, solve the equation $f(x) = 0$.'

This is the easy(ish) bit — just substitute for m and factorise.

If $m = 10$, you've got: this will factorise:

$$x^2 + 10x + 25 = 0$$
$$(x + 5)(x + 5) = 0$$
$$x = -5$$

In fact they're both perfect squares, which you could've predicted, since you're after a single root.

For $m = -10$, it's almost the same:

$$x^2 - 10x + 25 = 0$$
$$(x - 5)(x - 5) = 0$$
$$x = 5$$

Simultaneous Equations — two for the price of one... joy of joy...

I don't reckon there's anything difficult here. No alarms and no surprises. The simultaneous equations bit should be a doddle. Then there's some very basic graph sketching. Then a quadratic equation to solve, which is just a joke. The last part's a bit harder. You need to remember that there are always two answers when you take the square root of something — a positive one and a negative one. But as long as you remember this, and realise what to do, i.e. substitute and use your answer to the previous part, you'll be fine. Still, it's all good practice. And don't worry — they do get harder.

Paper 1 Q4 — Geometry

4 **(i)** Find the coordinates of the point A, when A lies at the intersection of the lines l_1 and l_2, and when the equations of l_1 and l_2 respectively are:

$$x - y + 1 = 0 \quad \text{and} \quad 2x + y - 8 = 0.$$ **[3]**

The points B and C have coordinates $(6, -4)$ and $(-\frac{4}{3}, -\frac{1}{3})$ respectively, and D is the midpoint of AC.

(ii) Find the equation of the line BD in the form $ax + by + c = 0$, where a, b and c are integers. **[5]**

(iii) Show that the triangle ABD is a right-angled triangle. **[3]**

(i) Finding A is easy — it's just Simultaneous Equations...

'Find the coordinates of the point A, when A lies at the intersection of the lines l_1 and l_2, and when the equations of l_1 and l_2 respectively are: $x - y + 1 = 0$ and $2x + y - 8 = 0$.'

Best draw a quick sketch so you know what you're looking for:

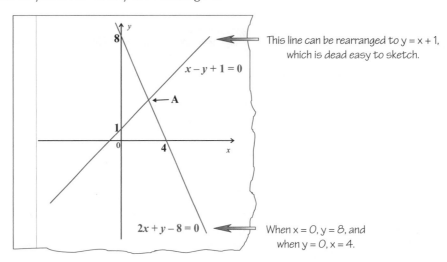

This line can be rearranged to y = x + 1, which is dead easy to sketch.

$x - y + 1 = 0$

A

$2x + y - 8 = 0$ — When x = 0, y = 8, and when y = 0, x = 4.

To find the coordinates of A, you need to solve the two lines as a pair of <u>simultaneous equations</u>.

Line 1 — l_1 — $x - y + 1 = 0$

Line 2 — l_2 — $2x + y - 8 = 0$

Forgotten everything you ever knew about simultaneous equations? Have a look at page 20.

Get rid of y to find x:

$l_1 + l_2$ $(x + 2x) + (-y + y) + (1 - 8) = 0$

$$3x - 7 = 0$$

$$x = \frac{7}{3}$$

Stick x=7/3 back into l_1 to find y:

l_1 $x - y + 1 = 0$

$$\frac{7}{3} - y + 1 = 0$$

$$y = \frac{7}{3} + 1 = \frac{10}{3}$$

So A is: $\left(\frac{7}{3}, \frac{10}{3}\right)$

Paper 1 Q4 — Geometry

(ii) | *Equation of a line — find the Gradient First...*

'Find the equation of the line BD in the form $ax + by + c = 0$...'

This question gives you loads of information. So draw a sketch.
Otherwise you won't have a clue what's going on. (Well I wouldn't anyway.)

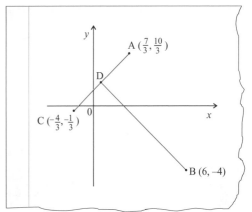

To find the equation of a line, you'll need its <u>gradient</u>.

And to find the gradient of BD, you'll need the <u>coordinates</u> of B and D
— you're given B in the question, but <u>you've got to find D</u>.

Find D (the midpoint of A and C)

You get the midpoint of two points by finding the <u>average</u>
of the x-coordinates, and the <u>average</u> of the y-coordinates.

Midpoint of AC is... $\left(\dfrac{x_A + x_C}{2}, \dfrac{y_A + y_C}{2} \right)$

x_A is the x-coordinate of the point A.
y_C is the y-coordinate of the point C.

which is... $= \left(\dfrac{\frac{7}{3} + \frac{-4}{3}}{2}, \dfrac{\frac{10}{3} + \frac{-1}{3}}{2} \right) = (\frac{1}{2}, \frac{3}{2})$

So D is... $(\frac{1}{2}, \frac{3}{2})$

Find the Gradient of BD

$\text{Gradient} = \dfrac{\text{difference in y - coordinates}}{\text{difference in x - coordinates}}$

m_{BD} is the gradient of the line BD.

$m_{BD} = \dfrac{y_D - y_B}{x_D - x_B} = \dfrac{\frac{3}{2} - -4}{\frac{1}{2} - 6} = \dfrac{3 + 8}{1 - 12} = -1$

So you've got the gradient... Well now you can do anything — you can sail around the world, you can become the richest person in the world, you can rule the world... you can become more powerful than you can possibly imagine...

Find the Equation of BD

I reckon y=mx+c is the nicest form for the equation of
a straight line. So I'd get it in that form first.

$y = m_{BD}x + c \Rightarrow y = -x + c$

The equation will be like this because
we've just worked out that the gradient
is –1 — you just need to find what c is.

Putting in the values for x and y at <u>either</u>
point B <u>or</u> point D will give you the value of c.

At point B, x=6 and y=-4

$y = mx + c$
$-4 = -6 + c$
$c = 2$

y=mx+c is great: m is the
gradient, and c is where the
line crosses the y-axis.

So equation for BD is: $y = -x + 2$

$x + y - 2 = 0$

Make sure it's in
the form the
question asks for.

Paper 1 Q4 — Geometry

(iii) *Right-angled triangle? Check if the lines are Perpendicular*

'Show that the triangle ABD is a right-angled triangle.'

The first thing to do is update your sketch (or do a new one) with the triangle ABD on.

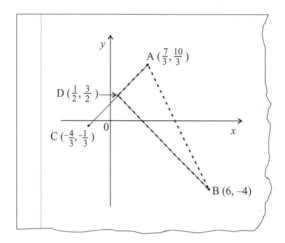

You've got to show that triangle ABD is right-angled. To do this, just show that two sides of the triangle are <u>perpendicular</u> to each other.
(That means you'll have a right angle in the corner where the sides meet.)

Hmm, what was that rule I used to know about perpendicular lines... ah yes, I remember:

> **Gradients of perpendicular lines multiply together to make –1.**

See page 27 for more info on perpendicular lines.

Your sketch should show you that the right angle will be at D, so you need to show that the lines **AD** and **BD** are perpendicular.

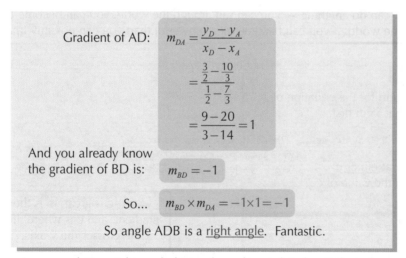

Gradient of AD: $m_{DA} = \dfrac{y_D - y_A}{x_D - x_A}$

$$= \dfrac{\frac{3}{2} - \frac{10}{3}}{\frac{1}{2} - \frac{7}{3}}$$

$$= \dfrac{9 - 20}{3 - 14} = 1$$

And you already know the gradient of BD is: $m_{BD} = -1$

So... $m_{BD} \times m_{DA} = -1 \times 1 = -1$

So angle ADB is a <u>right angle</u>. Fantastic.

Look how happy maths can make you...
I bet you're jumping for joy too...

So you've proved it's a right-angled triangle and completed a stinker of a question. Now you can...

...become the Master of the Universe...

To avoid getting into trouble while you're doing this question, you've got to draw what's happening before you do each part. So when you read stuff like, "The points B and C have coordinates (6, –4) and...", you should dive for your pencil and sketch all the information it gives you. And then <u>use</u> the sketch and <u>plan</u> how you're going to answer the question.

Paper 1 Q5 — Differentiation

> **5** Find *dy/dx* for each of the following:
> (i) $y = x^2$ [1]
> (ii) $y = 3x^4 - 2x$ [2]
> (iii) $y = (x^2 + 4)(x - 2)$ [2]

(i) Just use the Rule for Differentiation

'Find *dy/dx* for $y = x^2$.'

When you see *dy/dx* or f'(x), it means you need to **differentiate**.

If you've forgotten the rule for differentiating powers of x, flick back to page 33.

$$y = x^2$$
$$\frac{dy}{dx} = 2x^1$$
$$= \boxed{2x}$$

x¹ is just x.

(ii) Apply the rule to Each Term in Turn

'Find *dy/dx* for $y = 3x^4 - 2x$.'

Again you can use the rule for differentiation — but there are two terms, so you do it twice.

For the first term $n = 4$, and for the second term $n = 1$:

$$y = 3x^4 - 2x$$
$$\frac{dy}{dx} = 3(4x^{4-1}) - 2(1x^{1-1})$$
$$= 3(4x^3) - 2(1)$$
$$= \boxed{12x^3 - 2}$$

Don't let the coefficients in front of the x bits put you off — you just multiply by them. In maths-speak, that's d/dx (ky) = k dy/dx.

(iii) Expand first — then Differentiate

'Find *dy/dx* for $y = (x^2 + 4)(x - 2)$.'

When you need to differentiate an expression like this, you have to multiply it out first.

$$y = (x^2 + 4)(x - 2)$$
$$y = x^3 - 2x^2 + 4x - 8$$

This gives you a set of terms which are all powers of x. Which makes everything lovely.

Now differentiate each term:

$$y = x^3 - 2x^2 + 4x - 8$$
$$\frac{dy}{dx} = 3x^{3-1} - 2(2x^{2-1}) + 4(1x^{1-1}) - 0$$
$$= \boxed{3x^2 - 4x + 4}$$

Constant terms always disappear when you differentiate. That's because for a constant, the power of x is 0, so when you differentiate, you end up multiplying the term by 0, which gives 0. Glad we've cleared that up...

It'd be great if the summer term disappeared...

...and the spring one and the autumn one. Differentiation strikes fear into the hearts of the hardest of maths students. It can get a little sticky at times, so make sure you know the rules really well. Even then you probably won't be laughing, but at least you might be slightly more optimistic.

Paper 1 Q6 — Circles

6   The diagram shows a circle. A (2, 1) and B (0, –5) lie on the circle and AB is a diameter.
C (4, –1) is also on the circle.

(i)   Find the centre and radius of the circle.   [4]

(ii)   Show that the equation of the circle can be written in the form:
$$x^2 + y^2 - 2x + 4y - 5 = 0$$ [3]

(iii) The tangent at A and the normal at C cross at D.
Find the coordinates of D.   [8]

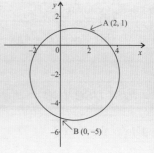

(i) | Average the Coordinates, then do a bit of Pythagoras

'Find the centre and radius of the circle.'

The centre of the circle must be at the midpoint of AB, since AB is a diameter.

To get the midpoint, you **average the x- and y-coordinates**.

Midpoint of AB is: $\left(\dfrac{2+0}{2}, \dfrac{1+-5}{2} \right) = (1, -2)$

As AB is the diameter, the length of the radius will be half the distance between A and B.

You can use Pythagoras' theorem to find the distance.
A quick sketch usually helps with these:

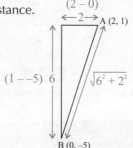

> If you've forgotten
> about surds, see p 3.

$$AB = \sqrt{40} = 2\sqrt{10}$$

Radius of the circle $= \dfrac{1}{2}AB = \sqrt{10}$

(ii) | Use the General Equation of a Circle

'Show that the equation of the circle can be written in the form: $x^2 + y^2 - 2x + 4y - 5 = 0$'

The general equation for a circle with centre (a, b) and radius r is: $(x - a)^2 + (y - b)^2 = r^2$

For this circle you have: $a = 1,\ b = -2,\ r = \sqrt{10}$

$(x - 1)^2 + (y + 2)^2 = 10$

If you multiply out the brackets, you should be able to get the form given in the question:

$(x - 1)(x - 1) + (y + 2)(y + 2) = 10$
$x^2 - 2x + 1 + y^2 + 4y + 4 = 10$
$x^2 + y^2 - 2x + 4y - 5 = 0$

Paper 1 Q6 — Circles

(iii) Tangents *touch* circles, Normals are at *right angles* to them

'Find the coordinates of D.'

A little bit of sketching on the diagram provided is always a good idea.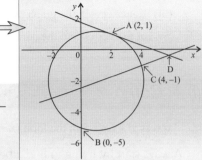

You need to find the equations of the tangent and normal and then work out where they cross.

The tangent at A: This is at right angles to the radius and diameter at A — you can use this to find the gradient.

The gradient of a line joining (x_1, y_1) to (x_2, y_2) is given by: $\dfrac{y_2 - y_1}{x_2 - x_1}$

For AB, gradient = $\dfrac{1--5}{2-0} = \dfrac{6}{2} = 3$

Use $y - y_1 = m(x - x_1)$ to get the equation of the tangent.

$m = -\dfrac{1}{3}$, (x_1, y_1) is the point (2, 1)

The **tangent is perpendicular to the diameter**, so use the gradient rule to find the gradient of the tangent.

$\dfrac{-1}{\text{gradient of AB}} = -\dfrac{1}{3}$

See p 27 if you've forgotten the gradient rule.

$y - 1 = -\dfrac{1}{3}(x - 2)$

$y - 1 = -\dfrac{1}{3}x + \dfrac{2}{3}$

$y = -\dfrac{1}{3}x + \dfrac{5}{3}$

So that's your equation for the tangent of the circle at point A.

The normal at C: A normal passes through the centre, so its gradient is the gradient from the centre to C.

You can use: $y - y_1 = m(x - x_1)$ to get the equation of the normal.

Gradient = $\dfrac{-1--2}{4-1} = \dfrac{1}{3}$

Where $m = \dfrac{1}{3}$, (x_1, y_1) is the point (4, –1).

$y--1 = \dfrac{1}{3}(x-4)$

$y + 1 = \dfrac{1}{3}x - \dfrac{4}{3}$

$y = \dfrac{1}{3}x - \dfrac{7}{3}$

So here's your equation for the normal at point C.

Where they cross: To get the coordinates of D, you need to find where the lines cross.

Put the equations equal to each other and solve:

$-\dfrac{1}{3}x + \dfrac{5}{3} = \dfrac{1}{3}x - \dfrac{7}{3}$

$\dfrac{5}{3} + \dfrac{7}{3} = \dfrac{1}{3}x + \dfrac{1}{3}x$

$4 = \dfrac{2}{3}x$

$x = 6$

Put this back into one of the equations to get y:

$y = -\dfrac{1}{3}x + \dfrac{5}{3}$

$y = -\dfrac{1}{3} \times 6 + \dfrac{5}{3}$

$y = -2 + \dfrac{5}{3}$

$y = -\dfrac{1}{3}$

So... D has coordinates $(6, -\dfrac{1}{3})$

Circles — a great shape for wheels...

OK, so at first glance this *looks* like a dirty great big question on circles. In fact, it's only half a dirty great big question on circles and half a dirty great big question on graphs. Tangents and normals questions are pretty much the same, whatever the shape of the curve you're looking at. All you need is the equation of the line and you're away. And if you notice, they give you the equation of the line in part (ii) — so you can still do part (iii), even if you make a pig's ear of the first two bits.

Paper 1 Q7 — Quadratics

7 **(i)** Express $x^2 - 6x + 5$ in the form $(x + a)^2 + b$. [2]

(ii) Factorise the expression $x^2 - 6x + 5$. [2]

(iii) Hence sketch the graph of $y = x^2 - 6x + 5$, clearly indicating the coordinates
of the vertex and the points where it cuts the axes. [2]

(i) | Complete the Square

'Express $x^2 - 6x + 5$ in the form $(x + a)^2 + b$.'

Always start by trying to find a because it's dead easy. You just divide the number in front of the x by two.

$\div 2$

In this case, half of -6 is -3: $x^2 - 6x + 5 = (x - 3)^2 + b$

Now multiply out... $(x - 3)^2 + b = x^2 - 6x + 9 + b$

then equate to find b: $x^2 - 6x + 5 = x^2 - 6x + 9 + b$

So: $5 = 9 + b$

$b = -4$

...then write it all out neatly — ta daaa:

$$x^2 - 6x + 5 = (x - 3)^2 - 4$$

(ii) | Oh Hallelujah — it's another Quadratic...

'Factorise the expression $x^2 - 6x + 5$.'

Write out the two brackets — the coefficient of x^2 is 1,
so start by writing x at the front of each bracket.

$x^2 - 6x + 5 = (x\qquad)(x\qquad)$

The numbers in the brackets have to multiply to make 5 and add together to make -6.

Oh look... -1 and -5 might just do it... $-1 + -5 = -6$ and $-1 \times -5 = 6$

$x^2 - 6x + 5 = (x - 1)(x - 5)$ Hurrah.

Quadratics are nice'n'easy to factorise when the number at the end is prime, because it only has one pair of factors.

(iii) | Find the Key Points Before you Sketch

'Hence sketch the graph of $y = x^2 - 6x + 5$, clearly indicating the coordinates of the vertex
and the points where it cuts the axes.'

"Hence" means you can use the work you've already done.

In part (i) you showed that $x^2 - 6x + 5 = (x - 3)^2 - 4$. You can use this to find the vertex.

$(x - 3)^2$ can't be less than zero (because it's squared, of course).

So the minimum must be when the bracket $= 0$, i.e. when: $x = 3$

and therefore when: $y = 0 - 4 = -4$

So the vertex of the graph is at $(3, -4)$.

The coefficient of x^2 is positive, so the graph is U-shaped, and you're looking for a minimum rather than a maximum.

Paper 1 Q7 — Quadratics

Now find where the graph crosses the two axes:

When the curve cuts the x-axis, y equals zero. So: $x^2 - 6x + 5 = 0$

Using your result from part (ii), you have: $(x - 1)(x - 5) = 0$

So $x = 1$ and $x = 5$, i.e. the graph cuts the x-axis at: $(1, 0)$ and $(5, 0)$

Two solutions, so your graph crosses the x-axis twice.

Some quadratics meet the x-axis just once, while others never have the pleasure.

When the curve cuts the y-axis, x equals zero. So: $y = 0^2 - (6 \times 0) + 5$

This gives: $y = 5$

So the graph cuts the y-axis at: $(0, 5)$

One solution, so your graph crosses the y-axis once. This is actually the case for all quadratics, but sometimes you can't see it happen.

They'll meet eventually...

Now you're ready for some Sketching Action

You know that a quadratic graph with a positive coefficient of x^2 will be a U-shape.
You now know the key points too, so plotting the graph is easy.

I always rotate the paper when I'm sketching quadratics so my hand is always in the right position to draw a <u>smooth curve</u>.

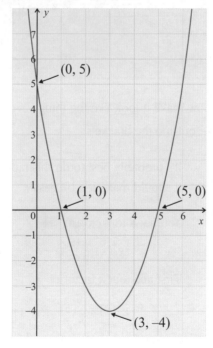

Knowing where the graph crosses the axes will help you decide the scale of the sketch.

Make sure you give yourself plenty of room to plot and label all the key points.

Another thing: check the entire curve is perfectly symmetrical about a vertical line that goes through the vertex. For example, you know the curve goes through $(0, 5)$, so it must go through $(6, 5)$ on the opposite side.

Are you up to it...

And your challenge is to sketch a quadratic graph — and I must warn that this one's an automatic lock-in.
First factorise, then find the vertices and intercepts, then sketch a smooth curve.
5 seconds... 4... 3... 2... 1... Get the crystal! Get the crystal!

Paper 1 Q8 — Differentiation

8 The function $g(x)$ is defined by $g(x) = \dfrac{x(3x^2 - 2x - 9)}{\sqrt{x}}$, for $x > 0$.

  (i) Show that the derivative of $g(x)$ can be written $g'(x) = \dfrac{3}{2\sqrt{x}}(5x^2 - 2x - 3)$. [3]

  (ii) Hence find the x-coordinate of the stationary point ($x > 0$) of $g(x)$. [3]

  (iii) By evaluating $g''(x)$ or otherwise, find the nature of this stationary point. [3]

A quick reminder before you start. It says in the question that $x > 0$. Don't forget this — they've said it for a <u>reason</u>.

(i) | Simplify it — then Differentiate...

'Show that the derivative of $g(x)$ can be written $g'(x) = \dfrac{3}{2\sqrt{x}}(5x^2 - 2x - 3)$.'

Not an easy start to the question — it looks like a bit of a nightmare in fact. But as long as you keep your head and remember the rules of differentiation, it won't really be that bad.

$$\frac{d}{dx}(x^n) = nx^{n-1}$$

First <u>rewrite</u> g(x) so that it's in a form you can differentiate (i.e. powers of x).

$$g(x) = \frac{x(3x^2 - 2x - 9)}{\sqrt{x}}$$
$$= \frac{3x^3 - 2x^2 - 9x}{x^{\frac{1}{2}}}$$
$$= 3x^{\frac{5}{2}} - 2x^{\frac{3}{2}} - 9x^{\frac{1}{2}}$$

Remember that $x^m \div x^n = x^{m-n}$

$$\frac{d}{dx}(x^n) = nx^{n-1}$$

Got total memory loss about everything to do with differentiation? Turn to page 33 for some help.

Then using the normal rule for differentiation...

$$g(x) = 3x^{\frac{5}{2}} - 2x^{\frac{3}{2}} - 9x^{\frac{1}{2}}$$
$$g'(x) = 3\left(\frac{5}{2}x^{\frac{3}{2}}\right) - 2\left(\frac{3}{2}x^{\frac{1}{2}}\right) - 9\left(\frac{1}{2}x^{-\frac{1}{2}}\right)$$
$$= \frac{15}{2}x^{\frac{3}{2}} - 3x^{\frac{1}{2}} - \frac{9}{2}x^{-\frac{1}{2}}$$

$g'(x) = \dfrac{dg}{dx}$

You've differentiated the function — now make it <u>look</u> like the thing in the <u>question</u> — with a factor of $\dfrac{3}{2\sqrt{x}}$ outside the brackets.

It's probably best to do this factorising in <u>two stages</u>...
It should make it a bit easier.

You need to factorise:

$$g'(x) = \frac{15}{2}x^{\frac{3}{2}} - 3x^{\frac{1}{2}} - \frac{9}{2}x^{-\frac{1}{2}}$$

It's already looking promising — you've got the 5, −2 and −3.

First take out a factor of $\frac{3}{2}$.

$$g'(x) = \frac{3}{2}\left(5x^{\frac{3}{2}} - 2x^{\frac{1}{2}} - 3x^{-\frac{1}{2}}\right)$$

Then take out a factor of $\dfrac{1}{\sqrt{x}}$. Don't panic... $\dfrac{1}{\sqrt{x}} = x^{-\frac{1}{2}}$

It's easy to make a mistake here, so you should definitely check that this is the same as what you had before by multiplying the brackets out.

$$g'(x) = \frac{3}{2}x^{-\frac{1}{2}}\left(5x^2 - 2x - 3\right)$$

Remember that when you multiply two numbers, you add the powers: $x^m x^n = x^{m+n}$

And then rewrite this to get it looking how the question wants it...

$$g'(x) = \frac{3}{2\sqrt{x}}\left(5x^2 - 2x - 3\right)$$

Paper 1 Q8 — Differentiation

(ii) Find **Stationary Points** by setting the derivative equal to **Zero**

'Hence find the x-coordinate of the stationary point ($x > 0$) of $g(x)$.'

The question uses the word 'hence', so you know you have to <u>use</u> the part of the question you've already answered.

To find a <u>stationary</u> point, set the derivative equal to <u>zero</u>. Remember — the derivative shows the gradient of the curve, so the curve is flat when the derivative is zero.

You need to solve $g'(x) = 0$

which is... $\dfrac{3}{2\sqrt{x}}(5x^2 - 2x - 3) = 0$

> You found this in the last part of the question.

...and because <u>anything times by zero is zero</u>, you can just work out when the quadratic bit is <u>zero</u>.

$$5x^2 - 2x - 3 = 0$$

Now, you can either factorise the quadratic or use the formula:

$$x = \frac{-b \pm \sqrt{b^2 - 4ac}}{2a}$$

> $5x^2 - 2x - 3 = 0$, so $a = 5$, $b = -2$ and $c = -3$

$$= \frac{2 \pm \sqrt{(-2)^2 - (4 \times 5 \times -3)}}{2 \times 5} = \frac{2 \pm \sqrt{4 + 60}}{10} = \frac{2 \pm 8}{10} = \frac{1 \pm 4}{5}$$

This gives the two possible answers as... $x = 1$ or $x = -\dfrac{3}{5}$

But since $x > 0$, only <u>one</u> of these is a possible solution, and that's... $x = 1$

> I said the fact that $x > 0$ would be important — that's why.

(iii) Differentiate **Again** to find the **Nature** of the stationary points

'...find the nature of this stationary point.'

Now you have to find the <u>nature</u> of this stationary point — which means deciding if it's a <u>minimum</u> or a <u>maximum</u>. To decide this, you have to differentiate again and find whether this second derivative is <u>positive</u> or <u>negative</u>.

$$g'(x) = \frac{15}{2}x^{\frac{3}{2}} - 3x^{\frac{1}{2}} - \frac{9}{2}x^{-\frac{1}{2}}$$

> Using the same rules as always to differentiate these powers of x.

$$g''(x) = \frac{15}{2}\left(\frac{3}{2}x^{\frac{1}{2}}\right) - 3\left(\frac{1}{2}x^{-\frac{1}{2}}\right) - \frac{9}{2}\left(-\frac{1}{2}x^{-\frac{3}{2}}\right)$$

$$= \frac{45}{4}x^{\frac{1}{2}} - \frac{3}{2}x^{-\frac{1}{2}} + \frac{9}{4}x^{-\frac{3}{2}}$$

$$\boxed{\frac{d}{dx}(x^n) = nx^{n-1}}$$

> Stick the value of x (= 1) into this equation.

$$g''(x) = \frac{45}{4}x^{\frac{1}{2}} - \frac{3}{2}x^{-\frac{1}{2}} + \frac{9}{4}x^{-\frac{3}{2}}$$

$$g''(1) = \left(\frac{45}{4} \times 1\right) - \left(\frac{3}{2} \times 1\right) + \left(\frac{9}{4} \times 1\right)$$

$$= \frac{45}{4} - \frac{3}{2} + \frac{9}{4} = \frac{45 - 6 + 9}{4} = \frac{48}{4} = 12$$

So the answer's <u>12</u> — but it's really only the fact that it's <u>positive</u> that's important.

Because that's <u>positive</u>, it means the stationary point is a <u>minimum</u>.

From Despair to Where? Differentiation — that's where...

This is quite a tricky question. Not only is the equation complicated, but it's pretty difficult to work out what the examiner actually wants you to do. And that differentiation looks a nightmare, what with all the fractions and minus signs floating about, but all you're doing is using the same rules that you'd always use. Just don't be tempted to try and save yourself a bit of time by rushing stuff like that — it never works. But really, this question is loads of simple stuff, all bunged together.

General Certificate of Education
Advanced Subsidiary (AS) and Advanced Level

Core 1 Mathematics — Practice Exam Two

You are NOT allowed to use a calculator.

---

1 (a) (i) Express $x^2 - 7x + 17$ in the form $(x - a)^2 + b$, where a and b are constants.

Hence state the maximum value of $f(x) = \dfrac{1}{x^2 - 7x + 17}$. [3]

(ii) Find the possible values of b if the equation $g(x) = 0$ is to have only one root,

where $g(x)$ is given by $g(x) = 3x^2 + bx + 12$. [3]

(b) (i) Factorise the quadratic expression $x^2 - 2x - 63$. [2]

(ii) Hence find all the solutions to the equation $f(x) = 0$, where $f(x)$ is given by $f(x) = x^4 - 2x^2 - 63$. [2]

(iii) Find the coordinates of all the stationary points of the graph $y = f(x)$, and find the gradient

at the point $x = 2$. [3]

---

2 A is the point $(8, -7)$ and B is the point $(2, -3)$.

(i) Find the equation of the line AB in the form: $ax + by + c = 0$ [3]

(ii) If B is the midpoint of AC, find the coordinates of C. [2]

The line L has equation $y = 2x + 13$.

(iii) Find the point where L crosses AC. [4]

---

3 A triangle has sides which lie on the lines given by the following equations:

AB: $y = 3$ BC: $2x - 3y - 21 = 0$ AC: $3x + 2y - 12 = 0$

(i) Find the coordinates of the vertices of the triangle. [5]

(ii) Show that the triangle is right-angled. [2]

The point D has coordinates $(3, d)$ and the point E has coordinates $(9, e)$.

(iii) If point D lies outside the triangle, but not on it, show that either $d > 3$ or $d < 1.5$. [2]

(iv) Given that E lies inside the triangle, but not on it, find the set of possible values for e. [3]

4 The circle with equation $x^2 - 6x + y^2 - 4y = 0$ crosses the y-axis at the origin and at the point A.

   (i) Find the coordinates of A.   [2]

   (ii) Rearrange the equation of the circle into the form: $(x - a)^2 + (y - b)^2 = c$ [4]

   (iii) Write down the radius and the coordinates of the centre of the circle.   [2]

   (iv) Find the equation of the tangent to the circle at A.   [4]

5 **(i)** Simplify $(\sqrt{3} + 1)(\sqrt{3} - 1)$.   [1]

   (ii) Rationalise the denominator of the expression $\dfrac{\sqrt{3}}{\sqrt{3} + 1}$.   [2]

   (iii) Simplify the expression: $\dfrac{x^3 \times x^4}{\sqrt{x^{10}}}$ [2]

6 **(i)** Find the set of solutions satisfying: $\sqrt{2}(x + 2\sqrt{2}) < 8 - \sqrt{2}x$ [3]

   (ii) Find the set of values of m that satisfy the inequality $x^2 - 6x + m > 0$ for all values of x.   [3]

   (iii) Solve the inequality: $-5 < x^2 + 4x - 5 < 7$ [6]

7 **(i)** Solve the equation: $\dfrac{4}{(x - 2)} = \dfrac{6}{(2x + 5)}$.   [3]

   (ii) Show that: $\dfrac{4}{9(x + 2)} + \dfrac{1}{3(x - 1)^2} + \dfrac{5}{9(x - 1)} = \dfrac{x^2}{(x + 2)(x - 1)^2}$.   [5]

8 This diagram shows the graph of $y = f(x)$.

Using a different set of axes for each one, sketch the following graphs.

Show as much of the graph as you can, labelling the axes wherever appropriate.

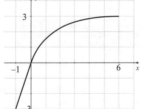

   (i) $y = f(x + 3)$ [2]

   (ii) $y = f(x) + 3$ [2]

   (iii) $y = f(-3x)$ [2]

   (iv) $y = 3f(x)$ [2]

Paper 2 Q1 — Quadratics

1 (a) (i) Express $x^2 - 7x + 17$ in the form $(x - a)^2 + b$, where a and b are constants.

Hence state the maximum value of $f(x) = \dfrac{1}{x^2 - 7x + 17}$. [3]

(ii) Find the possible values of b if the equation $g(x) = 0$ is to have only one root, where $g(x)$ is given by $g(x) = 3x^2 + bx + 12$. [3]

(b) (i) Factorise the quadratic expression $x^2 - 2x - 63$. [2]

(ii) Hence find all the solutions to the equation $f(x) = 0$, where $f(x)$ is given by $f(x) = x^4 - 2x^2 - 63$. [2]

(iii) Find the coordinates of all the stationary points of the graph $y = f(x)$, and find the gradient at the point $x = 2$. [3]

(a)(i) | An easy *Completing the Square* bit

'Express $x^2 - 7x + 17$ in the form $(x - a)^2 + b$, where a and b are constants.'

The question's just asking you to <u>complete the square</u>. The x^2 bit's got a coefficient of 1, so it's not so bad...

$$x^2 - 7x + 17$$

Rewrite it as one bracket squared + a number.

Don't forget — you just halve the coefficient of x to get the number in the brackets.

$$= \left(x - \frac{7}{2}\right)^2 + b$$

Find b by making the old and new equations equal to each other.

Have a look at pages 12 and 13 for more on completing the square.

$$x^2 - 7x + 17 = \left(x - \frac{7}{2}\right)^2 + b$$

$$x^2 - 7x + 17 = x^2 - 7x + \frac{49}{4} + b$$

Simplify the equation to get the value of b.

$$b = 17 - \frac{49}{4} = \frac{19}{4}$$

$$x^2 - 7x + 17 = \left(x - \frac{7}{2}\right)^2 + \frac{19}{4}$$

Finding the *Maximum*...

'Hence state the maximum value of $f(x) = \dfrac{1}{x^2 - 7x + 17}$.'

It'll be maximum when the quadratic in the denominator is as <u>small</u> as possible.

The minimum value of the denominator is easy to see using your answer to the first part — it's $\frac{19}{4}$, because the squared bit is never less than zero.

It says 'hence' — that's a pretty major clue that you've got to use the result from the first part to do this bit.

So...

$$f(x)_{max} = \frac{1}{\left(\frac{19}{4}\right)} = \frac{4}{19}$$

(a)(ii) | Use the *Discriminant*

'Find the possible values of b if the equation $g(x) = 0$ is to have only one root, where $g(x) = 3x^2 + bx + 12$.'

You can tell how many roots a function has got if you use the <u>quadratic formula</u>.

Think about $b^2 - 4ac$. If $b^2 - 4ac = 0$, then it's only got one root.

The $b^2 - 4ac$ bit is called the discriminant — and it's this part that tells you how many roots a quadratic has. See page 15.

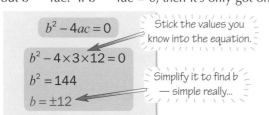

$$b^2 - 4ac = 0$$

Stick the values you know into the equation.

$$b^2 - 4 \times 3 \times 12 = 0$$

$$b^2 = 144$$

Simplify it to find b — simple really...

$$b = \pm 12$$

$$x = \frac{-b \pm \sqrt{b^2 - 4ac}}{2a}$$

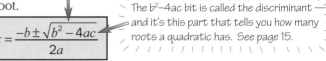

Examiners love adding things like g(x) — probably just to confuse you. They could have just said: 'Find the possible values of b if $3x^2 + bx + 12 = 0$.'

Paper 2 Q1 — Quadratics

(b)(i) | Factorising a quadratic... it must be Christmas

'Factorise the quadratic expression $x^2 - 2x - 63$'

The question tells you to use <u>factorisation</u> — so you know the numbers will be nice and easy.

$$x^2 - 2x - 63 = (x \quad)(x \quad)$$
$$= (x \quad 9)(x \quad 7)$$
$$= (x - 9)(x + 7)$$

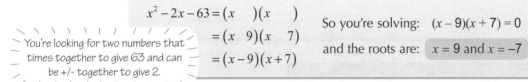
You're looking for two numbers that times together to give 63 and can be +/- together to give 2.

So you're solving: $(x - 9)(x + 7) = 0$

and the roots are: $x = 9$ and $x = -7$

For a reminder about how to factorise a quadratic — see pages 10-11.

(b)(ii) | A Quadratic in Disguise...

'Hence find all the solutions to the equation $f(x) = 0$, where $f(x)$ is given by $f(x) = x^4 - 2x^2 - 63$.'

The question uses the word 'hence' — so you're obviously going to have to use your answer to the part (i) to do this bit...

$f(x) = x^4 - 2x^2 - 63$...the new equation looks suspiciously like... $f(x) = x^2 - 2x - 63$

$f(x) = (x^2)^2 - 2(x^2) - 63$ ⟵ It's just got x^2's instead of x's.

Questions like this are covered on page 16.

So if you substitute $y = x^2$ in the new equation $f(y) = y^2 - 2y - 63$

From part (i), this has solutions $y = 9$ and $y = -7$

which means... $x^2 = 9$ or $x^2 = -7$

But since -7 is negative, the only possible solutions come from: $x^2 = 9$ i.e. $x = \pm 3$

(b)(iii) | A spot of differentiation... mmm, lovely Calculus...

'Find the coordinates of all the stationary points of the graph $y = f(x)$, and find the gradient at the point $x = 2$.'

Stationary points are easy — they're where the derivative (the gradient) has a value of zero.
So...

If $\dfrac{dy}{dx} = 4x^3 - 4x$, put this equal to nought, and factorise

This particular question's actually quite easy — $\frac{dy}{dx} = 0$ when $x = x^3$ so the solution should be obvious.

to get $0 = 4x(x^2 - 1)$
$$= 4x(x - 1)(x + 1)$$
$$\Rightarrow \text{when } \tfrac{dy}{dx} = 0, \ x = 0, \pm 1$$

Then you get the corresponding y-values by whacking these values of x into the original equation, $y = x^4 - 2x^2 - 63$.

So the stationary points are at
$(0, -63)$, $(1, -64)$ and $(-1, -64)$

Now find the gradient at $x = 2$. Just put $x = 2$ into the expression for the derivative, and hey presto...
$$\frac{dy}{dx} = 4x^3 - 4x = 4 \times 2^3 - 4 \times 2$$
$$= 32 - 8 = 24$$

And that's your answer. Smashing.

The square is complete..... Now I am the Master...

One of the commonest completing-the-square mistakes is not noticing when there's a number in front of the x^2 bit. Either that or noticing it but trying to stick it inside the brackets. Uh-uh. You've got to stick it outside, so it multiplies both the x^2 and x parts. And with the quadratic formula — you've got to be oh, so careful with those minus signs. So watch it.

Paper 2 Q2 — Equation of a Line

2 A is the point (8, –7) and B is the point (2, –3).

   (i) Find the equation of the line AB in the form: $ax + by + c = 0$ [3]

   (ii) If B is the midpoint of AC, find the coordinates of C.   [2]

   The line L has equation $y = 2x + 13$.

   (iii) Find the point where L crosses AC.   [4]

(i) *Equation of a line — find the Gradient first*

'Find the equation of the line AB in the form: $ax + by + c = 0$'

A sketch is bound to keep you on the right lines (so to speak).

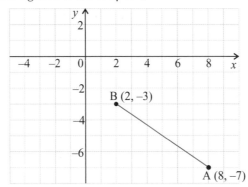

First find the gradient. I know there's no mention of *m* in the question, but once you've got an equation in the form $y = mx + c$, you'll be able to rearrange it into the form they're after.

$$\text{Gradient} = \frac{\text{difference in } y\text{-coordinates}}{\text{difference in } x\text{-coordinates}}$$

For AB: $\dfrac{-3 - -7}{2 - 8} = \dfrac{4}{-6} = -\dfrac{2}{3}$

Alternatively... you might like to find the gradient by using the graph. The gradient is 'rise over tread':

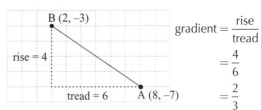

$$\text{gradient} = \frac{\text{rise}}{\text{tread}}$$
$$= \frac{4}{6}$$
$$= \frac{2}{3}$$

But it's sloping downwards, so the gradient will be negative: $-\dfrac{2}{3}$

The next step is to use: $y - y_1 = m(x - x_1)$

m is the gradient and (x_1, y_1) is a point on the line — you could use A or B.

$$y - y_1 = m(x - x_1)$$

Using B: $y - -3 = -\dfrac{2}{3}(x - 2)$

$$y + 3 = -\frac{2}{3}x + \frac{4}{3}$$

multiply through by 3 to remove the fractions: $3y + 9 = -2x + 4$

$$2x + 3y + 5 = 0$$

Paper 2 Q2 — Equation of a Line

(ii) | Read this one carefully

'If B is the midpoint of AC, find the coordinates of C.'

Usually you would be asked to find a **midpoint**, but in this question you need to find the **end** of the line.

It's a good idea to add C to your sketch. Remember B is the midpoint of AC.

You can suss out the position of C by counting along.
From A to B is **6 left** and **4 up**, so do the same again
from B to get C.

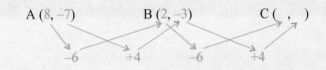

A (8, –7)   B (2, –3)   C (,)

–6   +4   –6   +4

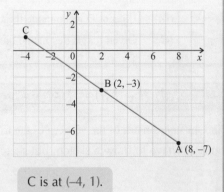

C is at (–4, 1).

(iii) | Intersection of Lines means Simultaneous Equations

'Find the point where L crosses AC.'

You want to find the point where $y = 2x + 13$ and $2x + 3y + 5 = 0$ meet. ⟵ The equation of the line AC is the same as for the line AB, which we found in part (i).

It is best to have them looking the same, so rearrange: $y = 2x + 13$
to give: $2x - y + 13 = 0$

So our simultaneous equations are:
$$2x + 3y + 5 = 0$$
$$2x - y + 13 = 0$$

Both equations have a '2x', so **subtracting bottom from top** will get rid of y.

$$2x + 3y + 5 = 0$$
$$- \quad 2x - y + 13 = 0$$
$$\overline{0x + 4y - 8 = 0}$$ ⟹
$$4y - 8 = 0$$
$$4y = 8$$
$$y = 2$$

Substitute back into one of the equations to get x:
$$2x - y + 13 = 0$$
$$2x - 2 + 13 = 0$$
$$2x + 11 = 0$$
$$2x = -11$$
$$x = -5.5$$

So the lines meet at: **(–5.5, 2)**

You should check this by trying it in the other equation:
$$2x + 3y + 5 = 0$$
$$2 \times -5.5 + 3 \times 2 + 5 = 0$$
$$-11 + 6 + 5 = 0$$

Hey presto, it works.

Is it me or are simultaneous equations just a bit dull...
Urgh... bring on the summer...

Paper 2 Q3 — Simultaneous Equations

> **3** A triangle has sides which lie on the lines given by the following equations:
>
> AB: $y = 3$ BC: $2x - 3y - 21 = 0$ AC: $3x + 2y - 12 = 0$
>
> **(i)** Find the coordinates of the vertices of the triangle. [5]
>
> **(ii)** Show that the triangle is right-angled. [2]
>
> The point D has coordinates $(3, d)$ and the point E has coordinates $(9, e)$.
>
> **(iii)** If point D lies outside the triangle, but not on it, show that either $d > 3$ or $d < 1.5$. [2]
>
> **(iv)** Given that E lies inside the triangle, but not on it, find the set of possible values for e. [3]

(i) Sketch the Lines before you deal with the Simultaneous Equations

'Find the coordinates of the vertices of the triangle.'

You don't *have* to draw a sketch, but sketches always help make questions like this clearer.

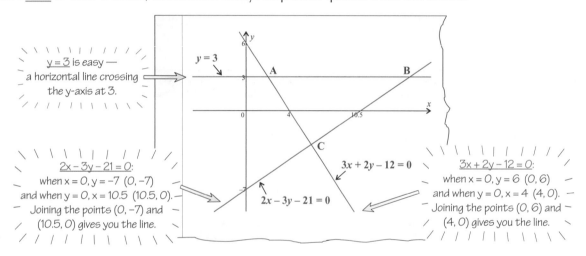

$y = 3$ is easy — a horizontal line crossing the y-axis at 3.

$2x - 3y - 21 = 0$:
when $x = 0$, $y = -7$ $(0, -7)$
and when $y = 0$, $x = 10.5$ $(10.5, 0)$.
Joining the points $(0, -7)$ and $(10.5, 0)$ gives you the line.

$3x + 2y - 12 = 0$:
when $x = 0$, $y = 6$ $(0, 6)$
and when $y = 0$, $x = 4$ $(4, 0)$.
Joining the points $(0, 6)$ and $(4, 0)$ gives you the line.

From the graph, you can see roughly where points A, B and C are, but you'll need to go through the algebra.

Finding A and B is straightforward, since you know that the y-coordinate is 3:

So for A:
$3x + 2y - 12 = 0$
$3x + (2 \times 3) - 12 = 0$
$3x + 6 - 12 = 0$
$3x = 6$
$x = 2$ A has coordinates $(2, 3)$.

And for B:
$2x - 3y - 21 = 0$
$2x - 3 \times 3 - 21 = 0$
$2x - 9 - 21 = 0$
$2x = 30$
$x = 15$ B has coordinates $(15, 3)$.

At C, you've got to solve a pair of simultaneous equations — bad luck. Label the equations:
$2x - 3y - 21 = 0$ (1)
$3x + 2y - 12 = 0$ (2)

Multiply equation (1) by 2 and equation (2) by 3 to get the coefficients of y equal:
$4x - 6y - 42 = 0$
$9x + 6y - 36 = 0$

Now add together to get rid of y:
$13x - 78 = 0$
$13x = 78$
$x = 6$

Substitute into equation (1) or (2) to find y:
Using eq. (2):
$3x + 2y - 12 = 0$ (2)
$(3 \times 6) + 2y - 12 = 0$
$18 + 2y - 12 = 0$
$2y + 6 = 0$
$2y = -6$
$y = -3$ C has coordinates $(6, -3)$.

These answers seem to make sense if you look back at the sketch, which is reassuring.

You should really check that works, by substituting your values into equation (1), but I'm running out of room.

Paper 2 Q3 — Simultaneous Equations

(ii) | A Right Angle probably means you need the Gradient Rule

'Show that the triangle is right-angled.'

Whenever you see 'right angle' in a coordinate geometry question, remember that the <u>gradients of two perpendicular lines multiply together to make –1</u>. So if you can show that they do, you've shown the triangle must be right-angled.

To find the gradient of each line, rearrange it into the form $y = mx + c$ — and m will be the gradient. You can tell from the diagram that the right angle looks like it's at C, so use BC and AC.

Line BC: $2x - 3y - 21 = 0$
$3y = 2x - 21$
divide by 3: $y = \frac{2}{3}x - 7$

Line AC: $3x + 2y - 12 = 0$
$2y = -3x + 12$
divide by 2: $y = -\frac{3}{2}x + 6$

⟹ You can see from the $y = mx + c$ form of the equations that the gradients are $\frac{2}{3}$ and $-\frac{3}{2}$.

$\frac{2}{3} \times -\frac{3}{2} = -1$ so the lines that meet at C are perpendicular, and the triangle must be right-angled.

(iii) | You know what I'm gonna say — sketch it on your graph

'If point D lies outside the triangle, but not on it, show that either $d > 3$ or $d < 1.5$.'

Point D has co-ordinates (3, d), so it must lie on the line x = 3. It lies outside the triangle, so you can see from the sketch that it must like **above the line AB** and **below the line AC**.

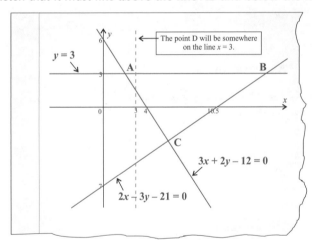

The point D will be somewhere on the line x = 3.

First off you can happily say that $d > 3$ is one condition, because $y = 3$ is the top of the triangle.

Secondly you have to work out the values of d that would give D is below the line AC.

If D was on the line AC, then (3, d) would satisfy $3x + 2y - 12 = 0$:

$3 \times 3 + 2d - 12 = 0$
$9 + 2d - 12 = 0$
$2d - 3 = 0$
$d = 1.5$

But as D can't be on the triangle, you need the y-coordinate to be less than 1.5, i.e. $d < 1.5$.

So the conditions are $d > 3$ or $d < 1.5$.

(iv) | Same thing again... almost

'Given that E lies inside the triangle, but not on it, find the set of possible values for e.'

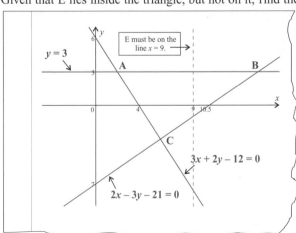

E must be on the line x = 9.

From the diagram you can tell that E is below AB and above BC. So, clearly, $e < 3$. Now work out the values of e so that E is above the line BC.

If e was on the line BC, then (9, e) would satisfy $2x - 3y - 21 = 0$:

$2x - 3y - 21 = 0$
$2 \times 9 - 3e - 21 = 0$
$18 - 3e - 21 = 0$
$-3e - 3 = 0$
$e = -1$

So above the line $e > -1$

Putting both conditions together gives you: $-1 < e < 3$

Simultaneous equations and graphs and gradients and inequalities...

These are a few of my favourite things.

Paper 2 Q4 — Circles

4   The circle with equation $x^2 - 6x + y^2 - 4y = 0$ crosses the y-axis at the origin and at the point A.

     (i)   Find the coordinates of A.   [2]

     (ii)   Rearrange the equation of the circle into the form: $(x - a)^2 + (y - b)^2 = c$ [4]

     (iii) Write down the radius and the coordinates of the centre of the circle.   [2]

     (iv) Find the equation of the tangent to the circle at A.   [4]

(i) | An Easy Start

'Find the coordinates of A.'

A is on the y-axis, so the x-coordinate is 0. Just put $x = 0$ into the equation and solve for y.

Here's the equation the question gives you: $x^2 - 6x + y^2 - 4y = 0$

Putting in $x = 0$: $0^2 - 6 \times 0 + y^2 - 4y = 0$

$$y^2 - 4y = 0$$

This will factorise: $y(y - 4) = 0$

$$y = 0 \text{ or } 4$$

$y = 0$ is the origin, so at A, $y = 4$.   A is at $(0, 4)$.

(ii) | This is like Completing the Square

'Rearrange the equation of the circle into the form: $(x - a)^2 + (y - b)^2 = c$'

The first thing to do is to find the two terms in the brackets — a and b.

You're basically completing the square for x and y separately, so you have to **halve the number in front of x to get a**, and **halve the number in front of y to get b**.

But for each one you'll end up with a number that you don't want (— namely a^2 and b^2). You'll need to take it away each time.

So here's the equation you're starting from: $x^2 - 6x + y^2 - 4y = 0$

Half of −6 is −3.

First deal with x: $x^2 - 6x = (x - 3)^2 - 9$ $(-3)^2 = 9$, so you need to take 9 away.

Half of −4 is −2.

Now the same for y: $y^2 - 4y = (y - 2)^2 - 4$ $(-2)^2 = 4$, so you need to take 4 away.

You can now put these new expressions back into the original equation.

$$(x - 3)^2 - 9 + (y - 2)^2 - 4 = 0$$
$$(x - 3)^2 + (y - 2)^2 - 13 = 0$$
$$(x - 3)^2 + (y - 2)^2 = 13$$

Paper 2 Q4 — Circles

(iii) What do **a**, **b** and **r** stand for?

'Write down the radius and the coordinates of the centre of the circle.'

When a question tells you to write something down, it means exactly that — no working's needed.
You do need to know some facts about the equation of a circle though.

> In the general equation for a circle $(x - a)^2 + (y - b)^2 = r^2$,
> the **centre** is (a, b) and the **radius** is r.

~ The only difference between this general equation of a circle
and the form that you found in part (ii) is the r^2 instead of c.
c must therefore equal r^2, so the radius (r) must be $\sqrt{c}$.

$a = 3$, $b = 2$, $r = \sqrt{c} = \sqrt{13}$

So the centre is $(3, 2)$, and the radius is $\sqrt{13}$.

(iv) A Tangent just touches the Circle

'Find the equation of the tangent to the circle at A.'

The tangent is at right angles to the radius. This is one of those mega-important facts about circles that you've
just gotta learn. You can find the gradient of the radius, and use that to find the gradient of the tangent.

The radius at A has a gradient of: $\dfrac{2-4}{3-0} = -\dfrac{2}{3}$

Remember the gradient rule? — Course you do.

$$\text{Gradient of line} = \frac{-1}{\text{gradient of perpendicular line}}$$

So the tangent at A has a gradient of: $\dfrac{-1}{-\frac{2}{3}} = \dfrac{3}{2}$

To find the equation of the tangent you can use the old favourite: $\boxed{y - y_1 = m(x - x_1)}$

m is the gradient $\left(\dfrac{3}{2}\right)$, and (x_1, y_1) is the coordinate of a known point on the line — in this case A $(0, 4)$.

$$y - y_1 = m(x - x_1)$$
$$y - 4 = \frac{3}{2}(x - 0)$$
$$y - 4 = \frac{3}{2}x$$
$$y = \frac{3}{2}x + 4$$

Circles — very popular in late 60s fashion design...

You have to make sure you're totally OK with completing the square before you attempt part (ii) or you'll just get horribly
confused. If you need to, go back to Section 2 and read up on completing the square, then come back to this one. This is
a pretty tough question actually (soz like) so if you can do all this stuff with ease, you're pretty much on top of things.

Paper 2 Q5 — Surds and Indices

5 **(i)** Simplify $(\sqrt{3}+1)(\sqrt{3}-1)$. [1]

(ii) Rationalise the denominator of the expression $\dfrac{\sqrt{3}}{\sqrt{3}+1}$. [2]

(iii) Simplify the expression: $\dfrac{x^3 \times x^4}{\sqrt{x^{10}}}$ [2]

(i) Just Multiply it out

'Simplify $(\sqrt{3}+1)(\sqrt{3}-1)$.'

Just multiply it out like you would if it was a quadratic:

$$(\sqrt{3}+1)(\sqrt{3}-1) = \sqrt{3}\times\sqrt{3} - 1\times\sqrt{3} + 1\times\sqrt{3} - 1\times 1$$
$$= 3 - \sqrt{3} + \sqrt{3} - 1$$
$$= 2$$

Rules of Surds

There's not really very much to remember.

$$\sqrt{ab} = \sqrt{a}\sqrt{b}$$

$$\sqrt{\frac{a}{b}} = \frac{\sqrt{a}}{\sqrt{b}}$$

$$a = (\sqrt{a})^2 = \sqrt{a}\sqrt{a}$$

Better still, you can save a bit of time if you recognise it as the **difference of 2 squares**:

$$(a+b)(a-b) = a^2 - b^2$$
$$(\sqrt{3}+1)(\sqrt{3}-1) = (\sqrt{3})^2 - 1^2$$
$$= 3 - 1$$
$$= 2$$

(ii) Get rid of the Square Root in the Denominator

'Rationalise the denominator of the expression $\dfrac{\sqrt{3}}{\sqrt{3}+1}$.'

To get rid of a root from the bottom of a fraction, you use the difference of two squares —

e.g. if the denominator's $\sqrt{a}+b$, you multiply by $\sqrt{a}-b$.

In this case, the denominator's $\sqrt{3}+1$, so you have to multiply by $\sqrt{3}-1$.

To get an equal fraction, you multiply the **top and bottom** of the fraction by the same amount.

$$\frac{\sqrt{3}}{\sqrt{3}+1} = \frac{\sqrt{3}(\sqrt{3}-1)}{(\sqrt{3}+1)(\sqrt{3}-1)} = \frac{3-\sqrt{3}}{2}$$

The bottom bit's just what you worked out for part (i).

Paper 2 Q5 — Surds and Indices

(iii) | *Work out the top and bottom — then Divide*

'Simplify the expression: $\dfrac{x^3 \times x^4}{\sqrt{x^{10}}}$,

As you would with any complicated fraction simplify the top first.
Here's a really important power law to remember:

$$y^m \times y^n = y^{m+n}$$ ⟹ When you've got a term raised to a power multiplied by the same term raised to another power, you can combine them by adding the powers.

So we have: $x^3 \times x^4 = x^{3+4} = x^7$

That's the top part sorted. Now for the bottom part. Remember the square root is the same as the power ½.

$$\sqrt{x^{10}} = (x^{10})^{\frac{1}{2}}$$

Use this rule of indices to get rid of the brackets.

Using: $$(y^m)^n = y^{mn}$$ ⟹ When you've got a bracket that's raised to a power, you can get rid of the power by multiplying the power outside the bracket by the power(s) inside the bracket.

$$(x^{10})^{\frac{1}{2}} = x^{10 \times \frac{1}{2}} = x^5$$

So you've got the top and bottom of the fraction. You've simplified the expression to: $\dfrac{x^7}{x^5}$.

Now you need to remember what happens when you divide powers...

$$y^m \div y^n = y^{m-n}$$ ⟹ When you've got a term raised to a power divided by the same term raised to another power, you can combine them by subtracting the powers.

So: $$\dfrac{x^7}{x^5} = x^{7-5} = x^2$$

Who'd have thought that something that looked so messy could end up being something so simple. Indices: an endless source of wonderment to us all.

To ignore surds would just be irrational...

Surds are really nice because they're clean — there are no messy decimals involved. They also save you the bother of using your calculator. Make sure you know the rules of surds, and then get plenty of practice at applying these rules — things like rationalising the denominator are very likely to come up in the exam. Indices are pretty darn useful too. If you know the three laws stated on this page, you should stand a decent chance in the exam.

Paper 2 Q6 — Inequalities

> 6   **(i)**   Find the set of solutions satisfying: $\sqrt{2}(x+2\sqrt{2}) < 8 - \sqrt{2}x$ [3]
>
>   **(ii)**   Find the set of values of m that satisfy the inequality $x^2 - 6x + m > 0$ for all values of x.   [3]
>
>   **(iii)**   Solve the inequality: $-5 < x^2 + 4x - 5 < 7$ [6]

(i) Just do it like a Normal Equation

'Find the set of solutions satisfying: $\sqrt{2}(x+2\sqrt{2}) < 8 - \sqrt{2}x$'

Treat it as if it were an equals sign in the middle. Get the xs on one side and the numbers on the other. This always works unless you have to divide or multiply through by a negative number (which changes the inequality round).

Start by multiplying out the bracket:

$$\sqrt{2}(x+2\sqrt{2}) < 8 - \sqrt{2}x$$
$$\sqrt{2}x + 2(\sqrt{2})^2 < 8 - \sqrt{2}x$$
$$\sqrt{2}x + 4 < 8 - \sqrt{2}x$$

Now get the numbers together and the xs together:

$$2\sqrt{2}x < 4$$

Now divide by $2\sqrt{2}$:

$$x < \frac{4}{2\sqrt{2}}$$
$$x < \frac{2}{\sqrt{2}}$$
$$x < \sqrt{2}$$

You have to rationalise the denominator here. Multiply top and bottom by root 2, simplify the bottom, then cancel: $\dfrac{2\sqrt{2}}{\sqrt{2}\sqrt{2}} = \dfrac{2\sqrt{2}}{2} = \sqrt{2}$

(ii) Use the Discriminant

'Find the set of values of m that satisfy the inequality $x^2 - 6x + m > 0$ for all values of x.'

If $x^2 - 6x + m > 0$, then there are no roots for the equation $x^2 - 6x + m = 0$.

If a quadratic equation has no real roots, then the discriminant must be less than zero (i.e. $b^2 - 4ac < 0$).

Remember — the discriminant is the bit from the quadratic formula inside the square root sign. So if the discriminant is less than 0, you'd have to take a square root of a negative number, which you can't do.

In this question, we have: $a = 1,\quad b = -6,\quad c = m$

$$b^2 - 4ac < 0$$
$$(-6)^2 - 4 \times 1 \times m < 0$$
$$36 - 4m < 0$$
$$36 < 4m$$
$$9 < m, \text{ or } \quad m > 9$$

Paper 2 Q6 — Inequalities

(iii) Do the inequality as two separate parts

'Solve the inequality: $-5 < x^2 + 4x - 5 < 7$'

This can be read as two separate inequalities: $-5 < x^2 + 4x - 5$ and $x^2 + 4x - 5 < 7$

The easiest way to tackle these inequalities is by drawing graphs. You need to turn the inequalities into equations which have an expression involving x, that is equal to 0. This should tell you where they cut the x-axis.

Taking the first inequality: $-5 < x^2 + 4x - 5$
$$x^2 + 4x > 0$$
$$x(x + 4) > 0$$
The equation will be: $x(x + 4) = 0$

$$x = 0 \text{ and } x = -4$$

And the second inequality: $x^2 + 4x - 5 < 7$
$$x^2 + 4x - 12 < 0$$
$$(x + 6)(x - 2) < 0$$
The equation will be: $(x + 6)(x - 2) = 0$

$$x = -6 \text{ and } x = 2$$

These are the points where the graphs cut the x-axis.

That's all the information you need to sketch the graphs. The coefficient of x^2 is positive so you know the graph will be u-shaped.

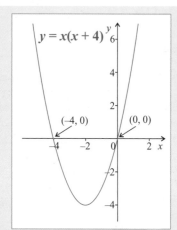

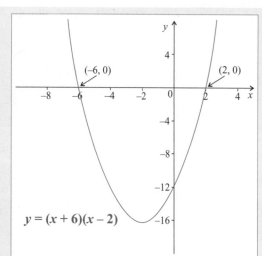

You can tell from the graph which regions satisfy the inequalities.

$$x(x + 4) > 0$$
So, $x < -4$ or $x > 0$
You want the part **above** the x-axis.

$$(x + 6)(x - 2) < 0$$
So, $-6 < x < 2$
You want the part **below** the x-axis.

You need the region where both inequalities are satisfied at the same time.
The two sets of answers can be put together to give:

$$-6 < x < -4 \text{ and } 0 < x < 2$$

Fraggle Rock — inequality in action...

A good example of inequality is the Doozer population of Fraggle Rock. Those poor souls build all day long, toiling in the darkness of a cave; then a fat greedy Fraggle eats their scaffolding without so much as a thank you. And yet still the Doozers are expected to remain cheerful. It makes my blood boil.

Paper 2 Q7 — Algebraic Fractions

7 **(i)** Solve the equation: $\dfrac{4}{(x-2)} = \dfrac{6}{(2x+5)}$. [3]

(ii) Show that: $\dfrac{4}{9(x+2)} + \dfrac{1}{3(x-1)^2} + \dfrac{5}{9(x-1)} = \dfrac{x^2}{(x+2)(x-1)^2}$. [5]

(i) | Get rid of Denominators when you can

'Solve the equation: $\dfrac{4}{(x-2)} = \dfrac{6}{(2x+5)}$.'

Whenever you have equations with fractions, it's a good idea to multiply each side of the equation by the denominators.

Here's what you've got to solve: $\dfrac{4}{(x-2)} = \dfrac{6}{(2x+5)}$

You've got **2** denominators, so multiply each side by $(x-2)$ **and** $(2x+5)$.

$$\frac{4(2x+5)(x-2)}{(x-2)} = \frac{6(2x+5)(x-2)}{(2x+5)}$$

$$4(2x+5) = 6(x-2)$$

This is cross-multiplication. Multiplying by the denominators of both sides will always get rid of the fractions.

Now multiply out and solve the equation:

$$4(2x+5) = 6(x-2)$$
$$8x + 20 = 6x - 12$$
$$2x = -32$$
$$x = -16$$

Check by subbing in this value of x into original equation:

$$\frac{4}{(-16-2)} = \frac{6}{(2\times(-16)+5)}$$

$$\frac{4}{-18} = \frac{6}{-27}$$

$$-\frac{2}{9} = -\frac{2}{9} \checkmark$$

(ii) | Adding fractions? — You need a Common Denominator

'Show that: $\dfrac{4}{9(x+2)} + \dfrac{1}{3(x-1)^2} + \dfrac{5}{9(x-1)} = \dfrac{x^2}{(x+2)(x-1)^2}$.'

This looks unlikely at first glance, but don't let that put you off. When you add algebraic fractions, the technique is the same as for normal fractions. You need a **common denominator**.

You basically need to multiply together **all** the denominators on the LHS.
But **when one term is a factor of another term**, you only need the **larger term**.

This only really makes sense when you see it work in practice...

Paper 2 Q7 — Algebraic Fractions

Finding the common denominator:

Look at the numbers in the LHS denominators. You've got 9, 3 and 9. Well, 3 is a factor of 9, so you only need the larger term, i.e. **9**.

The next thing to take a look at is the **(x + 2)**. This term **isn't** a factor of any of the other terms on the bottom, so you **will** need this.

Then you've got $(x - 1)^2$. One of the fractions has $(x - 1)$ as its denominator, which is of course a factor of $(x - 1)^2$. So again you just need the larger term, i.e. $(x - 1)^2$.

You've now looked at all the terms in the three denominators, and the common denominator you're left with is:

$$9(x + 2)(x - 1)^2$$

Make the denominator of each part the common denominator:

This involves multiplying the top and bottom of each fraction by **the terms of the common denominator that aren't in the denominator of the original fraction**. I know this sounds horrendous, but it'll make more sense when you see it in action.

Look at this fraction on the left: you've already got 9(x + 2) on the bottom, so you only need to multiply top and bottom by (x − 1)².

$$\frac{4}{9(x+2)} + \frac{1}{3(x-1)^2} + \frac{5}{9(x-1)}$$

You should be able to cancel down all the new fractions back to the original fractions.

E.g. $\dfrac{5(x+2)(x-1)}{9(x+2)(x-1)^2} = \dfrac{5}{9(x-1)}$

$$= \frac{4(x-1)^2}{9(x+2)(x-1)^2} + \frac{3(x+2)}{9(x+2)(x-1)^2} + \frac{5(x+2)(x-1)}{9(x+2)(x-1)^2}$$

Now multiply out the top bits.

$$= \frac{4x^2-8x+4}{9(x+2)(x-1)^2} + \frac{3x+6}{9(x+2)(x-1)^2} + \frac{5x^2+5x-10}{9(x+2)(x-1)^2}$$

Now you can turn it into just one fraction, and then it's just a question of collecting all the like terms.

$$= \frac{4x^2-8x+4}{9(x+2)(x-1)^2} + \frac{3x+6}{9(x+2)(x-1)^2} + \frac{5x^2+5x-10}{9(x+2)(x-1)^2}$$

$$= \frac{4x^2+5x^2-8x+3x+5x+4+6-10}{9(x+2)(x-1)^2}$$

$$= \frac{9x^2}{9(x+2)(x-1)^2}$$

The 9s cancel down:
$$= \frac{x^2}{(x+2)(x-1)^2}$$
⟸ and this is what the question asks for.

'The Denominator' — starring Armando Schwarzkopf...

Algebraic fractions — I shudder at the very sight of them. The most awkward bit when you're adding them, though, is finding the lowest common denominator. If it comes to the day and you can't remember how to do it, just multiply all the denominators together and use that. It's less elegant, but it *is* a common denominator.

Paper 2 Q8 — Graph Sketching

8 This diagram shows the graph of $y = f(x)$.

Using a different set of axes for each one, sketch the following graphs.
Show as much of the graph as you can, labelling the axes wherever appropriate.

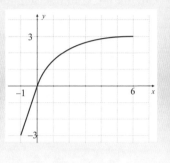

    (i) $y = f(x+3)$ [2]

    (ii) $y = f(x)+3$ [2]

    (iii) $y = f(-3x)$ [2]

    (iv) $y = 3f(x)$ [2]

There's not really too much to this question — but it looks a bit strange at first because there's hardly any numbers involved. To do this, you've really just got to know about the <u>transformations</u> in the question.

(i) Shift **Horizontally**...

When there's a number added or subtracted <u>inside</u> the brackets — the graph moves <u>sideways</u>.

$$y = f(x+3)$$

If the number is <u>added</u>, the graph moves to the <u>left</u>.
If the number is <u>subtracted</u>, the graph moves to the <u>right</u>.

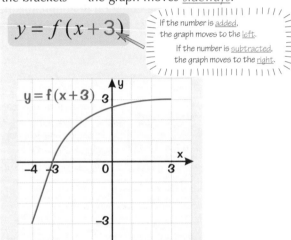

(ii) Shift **Vertically**...

When there's a number added or subtracted <u>outside</u> the brackets — the graph moves <u>up</u> or <u>down</u>.

$$y = f(x)+3$$

If the number is <u>added</u>, the graph moves <u>upwards</u>.
If the number is <u>subtracted</u>, the graph moves <u>downwards</u>.

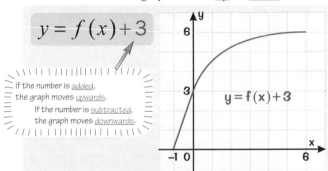

(iv) Stretch or squash **Vertically**...

When there's a number multiplied <u>outside</u> the brackets — the graph is stretched or squashed <u>vertically</u>.

$$y = 3f(x)$$

If the number is <u>bigger than 1</u>, the graph is <u>stretched</u>.
If the number is <u>smaller than 1</u>, the graph is <u>squashed</u>.
A negative number <u>reflects</u> the graph in the <u>x-axis</u>.

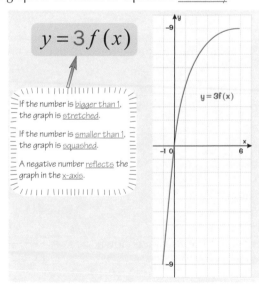

(iii) Stretch or squash **Horizontally**...

When there's a number multiplied <u>inside</u> the brackets — the graph is stretched or squashed <u>horizontally</u>.

$$y = f(-3x)$$

If the number is <u>bigger than 1</u>, the graph is <u>squashed</u>.
If the number is <u>smaller than 1</u>, the graph is <u>stretched</u>.
A <u>negative</u> number reflects the graph in the <u>y-axis</u>.

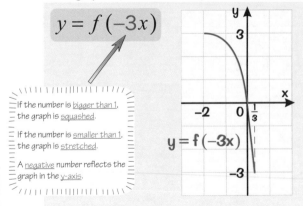

Could you be the most beautiful page in the world?

With questions like this, it's a very good idea to check your answer, especially if you're not sure. From the graph given in the question, you know that f(0)=0 and f(6)=3. So once you've drawn your transformed graph, you can check it's right. E.g. in part (a), my graph says when x=−3, y should be zero. Now: at x=−3, y=f(x+3)= f(−3+3)=f(0), which you know is 0. So it's right.

Four more things: 1) if something is altered <u>inside</u> the bracket, the graph is altered <u>horizontally</u>. 2) if something is changed <u>outside</u> the bracket, the graph is changed <u>vertically</u>. 3) <u>multiplying</u> stretches/squashes the graph. 4) <u>adding/subtracting</u> shifts the graph.

Answers

Section One — Algebra Basics

1) a) a & b are constants, x is a variable.

b) k is a constant, θ is a variable.

c) a, b & c are constants, y is a variable.

d) a is a constant, x, y are variables

2) Identity symbol is $\equiv$.

3) A, C and D are identities.

4) a) $-3/4$ b) f is undefined (or ∞)

c) $4/3$ d) 0

5) a) x^8 b) a^{15} c) x^6 d) a^8 e) x^4y^3z f) $\dfrac{b^2c^5}{a}$

6) a) 4 b) 2 c) 8 d) 1 e) $1/7$

7) a) $x = \pm\sqrt{5}$ b) $x = -2 \pm \sqrt{3}$

8) a) $2\sqrt{7}$ b) $\dfrac{\sqrt{5}}{6}$ c) $3\sqrt{2}$ d) $\dfrac{3}{4}$

9) a) $\dfrac{8}{\sqrt{2}} = \dfrac{8}{\sqrt{2}} \times \dfrac{\sqrt{2}}{\sqrt{2}} = \dfrac{8\sqrt{2}}{2} = 4\sqrt{2}$

b) $\dfrac{\sqrt{2}}{2} = \dfrac{\sqrt{2}}{\left(\sqrt{2}\right)^2} = \dfrac{1}{\sqrt{2}}$

(there are other possible ways to do these questions)

10) $136 + 24\sqrt{21}$

11) $3 - \sqrt{7}$

12) a) $a^2 - b^2$

b) $a^2 + 2ab + b^2$

c) $25y^2 + 210xy$

d) $3x^2 + 10xy + 3y^2 + 13x + 23y + 14$

13) a) $xy(2x + a + 2y\sin x)$

b) $\sin^2 x\left(1 + \cos^2 x\right)$

c) $8(2y + xy + 7x)$

d) $(x - 2)(x - 3)$

14) a) $\dfrac{52x + 5y}{60}$ b) $\dfrac{5x - 2y}{x^2 y^2}$

c) $\dfrac{x^3 + x^2 - y^2 + xy^2}{x\left(x^2 - y^2\right)}$

15) a) $\dfrac{3a}{2b}$ b) $\dfrac{2(p^2 + q^2)}{p^2 - q^2}$ c)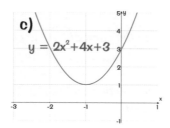

Section Two — Quadratics and Polynomials

1) a) $(x + 1)^2$ b) $(x - 10)(x - 3)$ c) $(x + 2)(x - 2)$

d) $(3 - x)(x + 1)$

e) $(2x + 1)(x - 4)$ f) $(5x - 3)(x + 2)$

2) a) $(x - 2)(x - 1) = 0$, so $x = 1$ or 2

b) $(x + 4)(x - 3) = 0$, so $x = -4$ or 3

c) $(2 - x)(x + 1) = 0$, so $x = 2$ or -1

d) $(x + 4)(x - 4) = 0$, so $x = 4$, -4

e) $(3x + 2)(x - 7) = 0$, so $x = -2/3$ or 7

f) $(2x + 1)(2x - 1) = 0$, so $x = \pm1/2$

g) $(2x - 3)(x - 1) = 0$, so $x = 1$ or $3/2$

3) a) $(x - 2)^2 - 7$; minimum value $= -7$ at $x = 2$, and this crosses the x-axis at $x = 2 \pm \sqrt{7}$

b) $\dfrac{21}{4} - \left(x + \dfrac{3}{2}\right)^2$; maximum value $= 21/4$ at $x = -3/2$, and this crosses the x-axis at

$$x = -\dfrac{3}{2} \pm \sqrt{\dfrac{21}{4}}$$

c) $2(x - 1)^2 + 9$; minimum value $= 9$ at $x = 1$, and this doesn't cross the x-axis.

d) $4\left(x - \dfrac{7}{2}\right)^2 - 1$; minimum value $= -1$ at $x = 7/2$, crosses the x-axis at $x = \dfrac{7}{2} \pm \dfrac{1}{2}$, i.e. $x = 3$ and $x = 4$.

4) a) $b^2 - 4ac = 16$, so 2 roots

b) $b^2 - 4ac = 0$, so 1 root

c) $b^2 - 4ac = -8$, so no roots

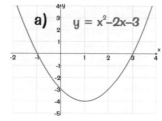

a) $y = x^2 - 2x - 3$

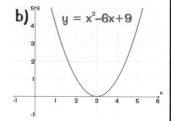

b) $y = x^2 - 6x + 9$

c) $y = 2x^2 + 4x + 3$

Answers

5) **a)** $x = \frac{7}{6} + \frac{\sqrt{13}}{6}$ or $x = \frac{7}{6} - \frac{\sqrt{13}}{6}$

   b) $x = \frac{3}{2} + \frac{\sqrt{13}}{2}$ or $x = \frac{3}{2} - \frac{\sqrt{13}}{2}$

   c) $x = -2 + \sqrt{10}$ or $x = -2 - \sqrt{10}$

6) $k^2 - (4 \times 1 \times 4) > 0$, so $k^2 > 16$ and so $k > 4$ or $k < -4$.

7) **a)** $2\sin^2 x - \sin x - 1 = 0$

$$\Rightarrow (2\sin x + 1)(\sin x - 1) = 0$$
$$\Rightarrow \sin x = -\frac{1}{2} \text{ or } \sin x = 1$$
$$\Rightarrow x = -30° \text{ or } 90°$$

   b) $\left(x^2 - 16\right)\left(x^2 - 1\right) = 0$

$$\Rightarrow x = \pm 4, \pm 1$$

   c) $\left(x^{\frac{2}{3}} - 1\right)\left(x^{\frac{2}{3}} - 4\right) = 0$

$$\Rightarrow x^{\frac{2}{3}} = 1, 4 \Rightarrow x = 1^{\frac{3}{2}}, 4^{\frac{3}{2}}$$
$$\Rightarrow x = \pm 1, \pm 8$$

Section Three — Simultaneous Equations and Inequalities

1) **a)** $x > -\frac{38}{5}$ **b)** $y \leq \frac{7}{8}$ **c)** $y \leq -\frac{3}{4}$

2) **(i)** $x > 5/2$ **(ii)** $x > -4$ **(iii)** $x \leq -3$

3) **a)** $-\frac{1}{3} \leq x \leq 2$ **b)** $x < 1 - \sqrt{3}$ or $x > 1 + \sqrt{3}$

   c) $x \leq -3$ or $x \geq -2$

4) **(i)** $x \leq -3$ and $x \geq 1$ **(ii)** $x < -1/2$ and $x > 1$
   (iii) $-3 < x < 2$

5) **a)** $(-3, -4)$ **b)** $\left(-\frac{1}{6}, -\frac{5}{12}\right)$

6) **a)** The lines meet at the points

$(2, -6)$ and $(7, 4)$.

   b) The line is a tangent to the parabola at the point

$(2, 26)$.

   c) The equations have no solution and so the line and the curve never meet.

7) **a)** $\left(\frac{1}{4}, -\frac{13}{4}\right)$ **b)** $(4, 5)$ **c)** $(-5, -2)$

Section Four — Coordinate Geometry and Graphs

1) **a)** **(i)** $y + 1 = 3(x - 2)$ **(ii)** $y = 3x - 7$
     (iii) $3x - y - 7 = 0$

   b) **(i)** $y + \frac{1}{3} = \frac{1}{5}x$ **(ii)** $y = \frac{1}{5}x - \frac{1}{3}$

     (iii) $3x - 15y - 5 = 0$

2) **a)** $\left(-\frac{1}{2}, 1\right)$ **b)** $\left(6, \frac{15}{2}\right)$ **c)** $\left(\frac{199}{2}, \frac{17}{2}\right)$

3) **a)** $6\sqrt{5}$ **b)** 10

4) **a)** $y = \frac{3}{2}x - 4$ **b)** $y = -\frac{1}{2}x + 4$

5) D is the point $\left(\frac{7}{2}, 7\right)$. So the line passing through

AD is $y = \frac{6}{5}x + \frac{14}{5} = \frac{1}{5}(6x + 14)$.

6) The midpoint of RS is $\left(5, \frac{13}{2}\right)$. The equation of the

required line is $y = \frac{8}{7}x + \frac{11}{14}$.

7) **a)** 3, (0, 0) **b)** 2, (2, −4) **c)** 5, (−3, 4)

8) **a)** 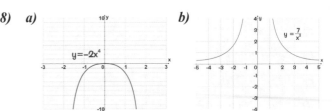 **b)**

   c) 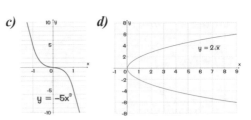 **d)**

   e)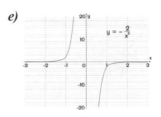

Answers

9)

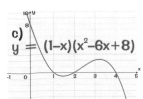
a) $y = (x-4)^3$

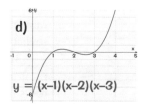
b) $y = (3-x)(x+2)^2$

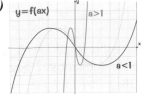
c) $y = (1-x)(x^2-6x+8)$

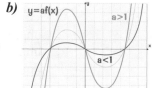
d) $y = (x-1)(x-2)(x-3)$

10)

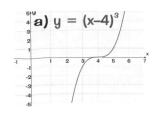
a) $y = f(ax)$, $a>1$, $a<1$

b) $y = af(x)$, $a>1$, $a<1$

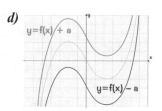
c) $y = f(x + a)$, $y = f(x - a)$

d) $y = f(x) + a$, $y = f(x) - a$

Section Five — Differentiation

1) $\dfrac{d}{dx}\left(x^n\right) = nx^{n-1}$

2) **a)** $\dfrac{dy}{dx} = 2x$, 2   **b)** $\dfrac{dy}{dx} = 4x^3 + \dfrac{1}{2\sqrt{x}}$, 9/2

    c) $\dfrac{dy}{dx} = -\dfrac{14}{x^3} + \dfrac{3}{2\sqrt{x^3}} + 36x^2$, 47/2

3) They're the same.

4) **a)** $\dfrac{dy}{dx} = 12x - 6$, so function is increasing when $x > 0.5$, and decreasing when $x < 0.5$.

    b) $\dfrac{dy}{dx} = -\dfrac{2}{x^3}$, so function is increasing when $x < 0$, and decreasing when $x > 0$.

5) A point where $\dfrac{dy}{dx} = 0$.

$y = x^3 - 9x^2 + 8x \Rightarrow \dfrac{dy}{dx} = 3x^2 - 18x + 8$, so stationary points are (5.52, –61.9) and (0.483, 1.88).

6) Differentiate again to find $\dfrac{d^2y}{dx^2}$. If this is positive, stationary point is a minimum; if it's negative, stationary point is a maximum.

7) $\dfrac{dy}{dx} = 3x^2 - \dfrac{3}{x^2}$; this is zero at (1, 4) and (–1, –4).

$\dfrac{d^2y}{dx^2} = 6x + \dfrac{6}{x^3}$; at x=1 this is positive, so (1, 4) is a minimum; at x=–1 this is negative, so (–1, –4) is a maximum.

8) Stationary points are (1, –2) (minimum) and (–1, 2) (maximum).

9) The tangent and normal must go through (16, 6).

Differentiate to find $\dfrac{dy}{dx} = \dfrac{3}{2}\sqrt{x} - 3$, so gradient at (16, 6) is 3. Therefore tangent can be written $y_T = 3x + c_T$; putting x=16 and y=6 gives $6 = 3\times16 + c_T$, so $c_T = -42$, and the equation of the tangent is $y_T = 3x - 42$.

The gradient of the normal must be $-\frac{1}{3}$ so the equation of the normal is $y_N = -\frac{1}{3}x + c_N$; substituting in the coordinates of the point (16, 6) give $6 = -\frac{16}{3} + c_N \Rightarrow c_N = \frac{34}{3}$; so the normal is $y = -\frac{1}{3}x + \frac{34}{3} = \frac{1}{3}(34 - x)$.

10) For both curves, when x=4, y=2, so they meet at (4,2). Differentiating the first curve gives $\dfrac{dy}{dx} = x^2 - 4x - 4$, which at x=4 is equal to –4.

Differentiating the other curve gives $\dfrac{dy}{dx} = \dfrac{1}{2\sqrt{x}}$, and so the gradient at (4, 2) is ¼. If you multiply these two gradients together you get –1, so the two curves are perpendicular at x=4.

Index

MRC151